Digitalization, Technologies, New Approaches, and Telemedicine in Dentistry and Craniofacial/ Temporomandibular Disorders

Digitalization, Technologies, New Approaches, and Telemedicine in Dentistry and Craniofacial/ Temporomandibular Disorders

Editors

Giuseppe Minervini
Rocco Franco

MDPI

Basel • Beijing • Wuhan • Barcelona • Belgrade • Novi Sad • Cluj • Manchester

Editors
Giuseppe Minervini
University of Campania
"Luigi Vanvitelli"
Naples
Italy

Rocco Franco
University of Rome, Tor Vergata
Rome
Italy

Editorial Office
MDPI AG
Grosspeteranlage 5
4052 Basel, Switzerland

This is a reprint of articles from the Special Issue published online in the open access journal *Applied Sciences* (ISSN 2076-3417) (available at: http://www.mdpi.com).

For citation purposes, cite each article independently as indicated on the article page online and as indicated below:

Lastname, A.A.; Lastname, B.B. Article Title. *Journal Name* **Year**, *Volume Number*, Page Range.

ISBN 978-3-7258-2217-1 (Hbk)
ISBN 978-3-7258-2218-8 (PDF)
doi.org/10.3390/books978-3-7258-2218-8

Contents

About the Editors

Giuseppe Minervini

Giuseppe Minervini graduated in Dental Medicine in July 2016 with honors. During his undergraduate studies, he participated in the Erasmus project at "Rey Juan Carlos Alcorcon" in Madrid, Spain, from September 2013 to June 2014. He received his Postgraduate Diploma in Orthodontics in December 2020 from the University of Campania, Luigi Vanvitelli, Naples, Italy, and later earned his PhD at the same institution (XXXIV cycle). In 2019, he attended the Tweed Study Course at the Charles H. Tweed International Foundation in Tucson, Arizona. Currently, he serves as Adjunct Professor at Saveetha Dental College and Hospitals, Saveetha University, Chennai.

Dr. Minervini is recognized as one of the top 10 experts worldwide in the field of "Temporomandibular disorders", according to Scopus rankings. He is a subject expert in dental materials and a tutor at the Orthodontics Dentistry School at the University of Campania, Luigi Vanvitelli. His contributions include serving as an editor for numerous dentistry journals (10 editorial boards and 14 Special Issues), and he is an active member of SIDO, EOS, and GSID. With over 188 publications, 70 posters, and an h-index of 32, he is a frequent speaker at national and international conferences. Dr. Minervini has received several awards for his contributions to the field. His research interests include biomedical and biomaterials applications in craniofacial, oral, and temporomandibular districts, as well as orthodontics, orofacial pain, temporomandibular joint disorders, and telemedicine.

Franco Rocco

Franco Rocco graduated in Dentistry and Dental Prosthetics at the University of Rome, Tor Vergata. He is currently in his third year of specialization in Oral Surgery (since November 2022). Dr. Rocco earned his PhD in Applied Medical–Surgical Science summa cum laude from the University of Rome, Tor Vergata. His academic journey reflects his deep commitment to advancing oral surgery techniques, with a special focus on dental prosthetics.

applied sciences

MDPI

Editorial

Digitalization, Technologies, New Approaches, and Telemedicine in Dentistry and Craniofacial/Temporomandibular Disorders

Rocco Franco [1] and Giuseppe Minervini [2,3,*]

[1] Department of Biomedicine and Prevention, University of Rome "Tor Vergata", 00100 Rome, Italy; rocco.franco@ptvonline.it
[2] Saveetha Dental College and Hospitals, Saveetha Institute of Medical and Technical Sciences (SIMATS), Saveetha University, Chennai 600001, India
[3] Multidisciplinary Department of Medical-Surgical and Odontostomatological Specialties, University of Campania "Luigi Vanvitelli", 80121 Naples, Italy
* Correspondence: giuseppe.minervini@unicampania.it

In recent years, the dental and craniofacial healthcare sectors have witnessed significant advancements driven by the integration of digitalization, new technologies, and telemedicine. These innovations have transformed diagnostic and therapeutic approaches, enhancing precision, efficiency, and patient outcomes. This editorial examines how these advancements are shaping dentistry and the management of craniofacial and temporomandibular disorders (TMDs), drawing insights from recent scholarly articles.

Digital dentistry has transformed the field, providing greater precision, efficiency, and improved patient outcomes [1–3]. Contemporary dental clinics often employ cutting-edge technologies, such as computer-aided design/computer-aided manufacturing (CAD/CAM), digital impressions, and 3D printing. CAD/CAM systems enable the precise fabrication of dental prostheses, ensuring better fitting and more aesthetically pleasing restorations compared to traditional methods [4]. Digital impressions taken with intraoral scanners have replaced the conventional and often uncomfortable impression materials, providing accurate data for prosthetic fabrication and reducing patient discomfort [5–7].

Furthermore, artificial intelligence (AI) is essential for improving diagnostic accuracy and treatment planning [8–10]. AI algorithms analyze large datasets, identify patterns, and make predictive assessments, facilitating early diagnosis and the creation of personalized treatment plans. For instance, AI can be used to detect caries, analyze radiographs, and predict the success of dental implants. The integration of AI with digital imaging tools such as cone-beam computed tomography (CBCT) and digital radiography enhances the precise diagnosis and treatment of complex dental conditions, including those affecting the craniofacial region [11–14].

Modern technological development has made a significant impact on the diagnostic and therapeutic approach to craniofacial disorders. Among the most optimistic ways is the application of transcutaneous electrical nerve stimulation, or TENS, a level-diagnostic and effective method of treating functional mandibular lateral deviation. Research shows that TENS can help visualize neuromuscular system trends, providing insights into mandibular position and enabling more accurate diagnoses [15]. This neuromuscular approach is crucial in addressing functional disorders that traditional diagnostic tools may overlook.

Furthermore, the implementation of 3D imaging and modeling has enhanced the understanding and treatment of craniofacial abnormalities. These technologies allow medical professionals to observe the intricate structure of the craniofacial area, strategize surgeries with enhanced accuracy, and make more precise predictions about the results after an operation [16]. This precision is particularly beneficial in orthognathic surgeries and the management of congenital craniofacial anomalies, ensuring better functional and aesthetic results for patients.

Citation: Franco, R.; Minervini, G. Digitalization, Technologies, New Approaches, and Telemedicine in Dentistry and Craniofacial/Temporomandibular Disorders. *Appl. Sci.* **2024**, *14*, 5871. https://doi.org/10.3390/app14135871

Received: 29 June 2024
Accepted: 3 July 2024
Published: 5 July 2024

Temporomandibular disorders (TMDs) are complex conditions that necessitate a multidisciplinary approach in order to achieve effective therapy [17]. Recent advancements have introduced novel diagnostic and therapeutic modalities that enhance the understanding and treatment of TMDs. Digital occlusal analysis systems, for example, provide detailed insights into bite force distribution and temporomandibular joint (TMJ) function, allowing for more targeted and effective treatments. Telemedicine has become a crucial instrument for treating temporomandibular disorders (TMDs), especially in light of the COVID-19 pandemic [18]. Teleconsultations enable patients to access specialized assistance and continuous therapy without the necessity of in-person appointments. This is particularly advantageous for individuals with limited mobility or those residing in isolated regions. Digital platforms enable the exchange of diagnostic data, such as digital scans and radiography, which enable clinicians to remotely evaluate and track patients' ailments.

Telemedicine is revolutionizing the management of TMDs and expanding access to dental care overall [19,20]. By using telehealth platforms, dentists can conduct virtual consultations, provide real-time treatment recommendations, and monitor patients' progress remotely. This approach is particularly advantageous for managing chronic conditions, post-operative care, and follow-up appointments. Additionally, telemedicine supports patient education and engagement by providing platforms for interactive communication. Patients can receive personalized oral hygiene instructions, dietary advice, and educational resources that empower them to take an active role in their oral health care [21].

The efficacy of elastodontic devices in managing TMDs has been a recent focus of research. A study by Ortu et al. compared the clinical efficacy of the Eptamed elastodontic device to a common bite sold in pharmacies, assessing surface electromyography (sEMG) and kinesiography activities of masticatory and postural muscles before and after six months of treatment [22]. The Eptamed device was found to significantly improve the electrical activity in the examined muscles, promoting greater relaxation and reducing the risk of worsening TMD symptoms compared to a standard bite, which showed a general worsening of muscle activity [22]. These findings indicate that elastodontic devices have the capacity to be an effective and minimally invasive therapy option for temporomandibular disorders (TMDs) as they assist in maintaining a proper equilibrium of the nerves and muscles involved, hence improving outcomes for patients.

The integration of digitalization, cutting-edge technologies, and telemedicine in dentistry and the treatment of craniofacial/TMD diseases is introducing a new period characterized by accuracy, effectiveness, and patient-focused healthcare. These innovations not only enhance diagnostic accuracy and treatment outcomes but also improve accessibility and convenience for patients. As these technologies evolve, their adoption is set to become more widespread, fundamentally transforming dental and craniofacial healthcare. The future of this field relies on the ongoing adoption of these cutting-edge innovations, ensuring patients receive the highest standard of care. Implementing these enhancements will lead to improved medical results, heightened patient contentment, and a more efficient healthcare system.

Conflicts of Interest: The authors declare no conflicts of interest.

References

1. Alqahtani, J.; Alhemaid, G.; Alqahtani, H.; Abughandar, A.; AlSaadi, R.; Algarni, I.; AlSharif, W.; Al-Harbi, S.; Burwaih, R.; Hasan, A.; et al. Digital Diagnostics and Orthodontic Practice. *J. Healthc. Sci.* **2022**, *2*, 112–117. [CrossRef]
2. Joda, T.; Ferrari, M.; Gallucci, G.O.; Wittneben, J.; Brägger, U. Digital Technology in Fixed Implant Prosthodontics. *Periodontology 2000* **2017**, *73*, 178–192. [CrossRef] [PubMed]
3. Khurshid, Z. Digital Dentistry: Transformation of Oral Health and Dental Education with Technology. *Eur. J. Dent.* **2023**, *17*, 943–944. [CrossRef] [PubMed]
4. Çin, V.; İzgi, A.D.; Kale, E.; Yilmaz, B. Marginal and Internal Fit of Monolithic Zirconia Crowns Fabricated by Using Two Different CAD-CAM Workflows: An In Vitro Study. *Prosthesis* **2023**, *5*, 35–47. [CrossRef]

5. Tallarico, M.; Cuccu, M.; Meloni, S.M.; Lumbau, A.I.; Baldoni, E.; Pisano, M.; Fiorillo, L.; Cervino, G. Digital Analysis of a Novel Impression Method Named the Biological-Oriented Digital Impression Technique: A Clinical Audit. *Prosthesis* **2023**, *5*, 992–1001. [CrossRef]
6. D'Ambrosio, F.; Giordano, F.; Sangiovanni, G.; Di Palo, M.P.; Amato, M. Conventional versus Digital Dental Impression Techniques: What Is the Future? An Umbrella Review. *Prosthesis* **2023**, *5*, 851–875. [CrossRef]
7. Yuzbasioglu, E.; Kurt, H.; Turunc, R.; Bilir, H. Comparison of Digital and Conventional Impression Techniques: Evaluation of Patients' Perception, Treatment Comfort, Effectiveness and Clinical Outcomes. *BMC Oral Health* **2014**, *14*, 10. [CrossRef]
8. Subramanian, A.K.; Chen, Y.; Almalki, A.; Sivamurthy, G.; Kafle, D. Cephalometric Analysis in Orthodontics Using Artificial Intelligence—A Comprehensive Review. *Biomed. Res. Int.* **2022**, *2022*, 1880113. [CrossRef] [PubMed]
9. Silva, T.P.; Hughes, M.M.; Menezes, L.D.S.; de Melo MD, F.B.; Freitas, P.H.L.D.; Takeshita, W.M. Artificial Intelligence-Based Cephalometric Landmark Annotation and Measurements According to Arnett's Analysis: Can We Trust a Bot to Do That? *Dentomaxillofac. Radiol.* **2022**, *51*, 20200548. [CrossRef]
10. Gokdeniz, S.T.; Kamburoğlu, K. Artificial Intelligence in Dentomaxillofacial Radiology. *World J. Radiol.* **2022**, *14*, 55–59. [CrossRef]
11. Duran, G.S.; Gökmen, Ş.; Topsakal, K.G.; Görgülü, S. Evaluation of the Accuracy of Fully Automatic Cephalometric Analysis Software with Artificial Intelligence Algorithm. *Orthod. Craniofacial Res.* **2023**, *26*, 481–490. [CrossRef]
12. Kılınç, D.D.; Kırcelli, B.H.; Sadry, S.; Karaman, A. Evaluation and Comparison of Smartphone Application Tracing, Web Based Artificial Intelligence Tracing and Conventional Hand Tracing Methods. *J. Stomatol. Oral Maxillofac. Surg.* **2022**, *123*, e906–e915. [CrossRef] [PubMed]
13. Görürgöz, C.; İçen, M.; Kurt, M.; Aksoy, S.; Bakırarar, B.; Rozylo-Kalinowska, I.; Orhan, K. Degenerative Changes of the Mandibular Condyle in Relation to the Temporomandibular Joint Space, Gender and Age: A Multicenter CBCT Study. *Dent. Med. Probl.* **2023**, *60*, 127–135. [CrossRef]
14. Lombardo, G.; Signoriello, A.; Marincola, M.; Bonfante, E.A.; Díaz-Caballero, A.; Tomizioli, N.; Pardo, A.; Zangani, A. Five-Year Follow-Up of 8 and 6 mm Locking-Taper Implants Treated with a Reconstructive Surgical Protocol for Peri-Implantitis: A Retrospective Evaluation. *Prosthesis* **2023**, *5*, 1322–1342. [CrossRef]
15. Ortu, E.; Di Nicolantonio, S.; Mummolo, A.; Cattaneo, R.; Pietropaoli, D.; Monaco, A. Use of Tens in the Diagnosis of Functional Mandibular Lateral Deviation. *Appl. Sci.* **2023**, *13*, 13258. [CrossRef]
16. Barayan, M.A.; Qawas, A.A.; Alghamdi, A.S.; Alkhallagi, T.S.; Al-Dabbagh, R.A.; Aldabbagh, G.A.; Linjawi, A.I. Effectiveness of Machine Learning in Assessing the Diagnostic Quality of Bitewing Radiographs. *Appl. Sci.* **2022**, *12*, 9588. [CrossRef]
17. Minervini, G.; Franco, R.; Marrapodi, M.M.; Di Blasio, M.; Isola, G.; Cicciù, M. Conservative Treatment of Temporomandibular Joint Condylar Fractures: A Systematic Review Conducted According to PRISMA Guidelines and the Cochrane Handbook for Systematic Reviews of Interventions. *J. Oral Rehabil.* **2023**, *50*, 886–893. [CrossRef]
18. Minervini, G.; Franco, R.; Marrapodi, M.M.; Almeida, L.E.; Ronsivalle, V.; Cicciù, M. Prevalence of Temporomandibular Disorders (TMD) in Obesity Patients: A Systematic Review and Meta-analysis. *J Oral Rehabil* **2023**, *50*, 1544–1553. [CrossRef] [PubMed]
19. Inchingolo, A.M.; Inchingolo, A.D.; Settanni, V.; De Leonardis, N.; Campanelli, M.; Garofoli, G.; Benagiano, S.; Malcangi, G.; Minetti, E.; Palermo, A.; et al. Correlation between Temporomandibular Disorders and Tinnitus and Possible Treatment Strategies: Comprehensive Review. *Appl. Sci.* **2023**, *13*, 8997. [CrossRef]
20. Minervini, G.; Franco, R.; Marrapodi, M.M.; Di Blasio, M.; Ronsivalle, V.; Cicciù, M. Children Oral Health and Parents Education Status: A Cross Sectional Study. *BMC Oral Health* **2023**, *23*, 787. [CrossRef]
21. Ceylan, G.; Eken, M.O.; Yuruk, S.; Emir, F. Examining the Influence of Self-Esteem and Digital Literacy on Professional Competence Factors in Dental Education: A Cross-Sectional Study. *Appl. Sci.* **2023**, *13*, 9411. [CrossRef]
22. Ortu, E.; Di Nicolantonio, S.; Cova, S.; Pietropaoli, D.; De Simone, L.; Monaco, A. Efficacy of Elastodontic Devices in Temporomandibular Disorder Reduction Assessed by Computer Aid Evaluation. *Appl. Sci.* **2024**, *14*, 1651. [CrossRef]

applied sciences

MDPI

Article

Efficacy of Elastodontic Devices in Temporomandibular Disorder Reduction Assessed by Computer Aid Evaluation

Eleonora Ortu [1], Sara Di Nicolantonio [1,*], Samuele Cova [2], Davide Pietropaoli [1], Lucia De Simone [1] and Annalisa Monaco [1]

[1] MeSVA Department, Dental Unit, University of L'Aquila, P.le S. Tommasi, 67100 L'Aquila, Italy; eleonora.ortu@univaq.it (E.O.); davide.pietropaoli@univaq.it (D.P.); lucia.desimone@graduate.univaq.it (L.D.S.); annalisa.monaco@univaq.it (A.M.)

[2] DDS, Freelance, Trento, Via Tiberio Claudio, 18, 38023 Cles, Italy; covasamuele@hotmail.com

* Correspondence: saradinicolantonio@libero.it

Abstract: Background: The main objective of this study was to compare the clinical efficacy of two different devices, the Eptamed elastodontic device and a common bite sold in pharmacies, by assessing a computer aid evaluation of patients' surface electromyography (sEMG) and kinesiography activity of four pairs of masticatory and postural muscles (anterior temporalis, digastricus, masseters and sternocleidomastoids muscles) before and after 6 months of treatment. Materials and Methods: Twelve adult patients with temporomandibular disorders and in need of orthodontic treatment were enrolled in the study and divided into cases and controls. Cases underwent orthodontic treatment with the Eptamed elastodontic device, while controls were treated with a bite sold in pharmacies. Both groups underwent electromyographic and kinesiographic examinations before and after 6 months from the start of treatment. Results: The Eptamed device was found to guarantee an improvement in the electrical activity of the muscles examined. The subjects in the control group, on the other hand, had a general worsening of electrical activity after wearing a splint purchased in a pharmacy. As for the kinesiographic examination, there was no significant improvement in both groups. Conclusions: the use of the Eptamed device in subjects with TMD ensured a greater relaxation of the chewing muscles than a standard bite, effectively reducing the risk of worsening the symptomatology of temporomandibular disorders.

Keywords: elastodontic device; bite standard; EMG; KNG; TMD(s)

Citation: Ortu, E.; Di Nicolantonio, S.; Cova, S.; Pietropaoli, D.; De Simone, L.; Monaco, A. Efficacy of Elastodontic Devices in Temporomandibular Disorder Reduction Assessed by Computer Aid Evaluation. *Appl. Sci.* **2024**, *14*, 1651. https://doi.org/10.3390/app14041651

Academic Editor: Bruno Chrcanovic

Received: 27 November 2023
Revised: 12 February 2024
Accepted: 15 February 2024
Published: 18 February 2024

1. Introduction

Temporomandibular disorder(s) (TMDs) is a term involving dysfunction and pain of the masticatory muscles, temporomandibular joints (TMJs) and surrounding tissues. It is a significant public health problem and represents the most common cause of nondental orofacial pain [1,2]. In the TMD sufferers, during palpation of the TMJ, in addition to hyperalgesia (usually detected by applying pressure at the level of the masticatory muscles or at the level of the TMJ), noises or clicks in the opening, closing or lateral movements of the jaw are also present. Actually, the chronic pain condition with which such individuals are forced to live can also have repercussions at the level of the psychological and social component of the patients [3]. To date, in fact, TMDs are considered bio-psycho-social disorders, in which the genesis and representation of symptoms are related to the interaction between objective biological aspects and the more complex aspects of the individual's psychological sphere and social interaction [4]. This is so much so that a recent systematic review of the literature [5] considered TMDs one of the factors that, more than all other oral conditions, can compromise the quality of life related to oral health (OHRQoL), so diagnosis and proper treatment is necessary to relieve pain and improve quality of life. TMD can occur at any age; in particular, 13% of subjects with temporomandibular disorders are elderly, from 13 to 30% are adults and 22% are teenagers [6].

Nowadays, the etiology of TMD remains a subject of great controversy and is generally viewed as multifactorial. The American Association of Oro-facial Pain considered orthodontic treatments managed in the absence of good neuromuscular balance to be among the major causes of temporomandibular disorders (in addition to microtrauma, macrotrauma and anatomical, hormonal and genetic factors) [7]. In general, the aim of an orthodontic treatment is to achieve functional and esthetic occlusion by using fixed or removable appliances of different types. In this way, in addition to changing the positioning of the dental elements and skeletal structures, the craniofacial structures can be oriented toward a correct direction of growth [8]. But, an additional goal of proper orthodontic treatment should be to achieve neuromuscular balance. The lack of neuromuscular balance after orthodontics, in fact, can induce the development of temporomandibular disorders (TMDs) [9].

To date, an increasing number of adults require orthodontic treatment; however, temporomandibular disorders are more common at this age. In fact, according to LeResche [10], pain associated with temporomandibular disorders occurs in 10% of the population over the age of 18 and is a condition that prevails in adolescents and middle-aged adults, while it is less common in children and elderly people. For this reason, orthodontic treatment could be risky, causing a worsening of symptoms in individuals with TMD or inducing the development of temporomandibular disorders in previously healthy subjects. The application of excessive orthodontic forces can cause occlusal trauma that can affect not only the TMJ but also periodontal tissues. In fact, according to studies by Pola et al., the presence of occlusal trauma is a risk factor for the progression of periodontal disease [11]. In addition, in subjects who already have periodontitis, it can exacerbate the inflammatory process of the dental support tissues with increased levels of RANKL in the GCF (gingival crevicular fluid) and tissue and bone damage [12]. It is, therefore, necessary during treatment to achieve tooth contacts that are stable and atraumatic, avoiding periodontal and joint repercussions. Therefore, a careful clinical examination with palpation of the TMJ and masticatory muscles during orthodontic treatment would be appropriate. Electromyographic and kinesiographic examination can be of fundamental help in this regard. They are used as diagnostic tools in the context of temporomandibular disorders [13]. EMG evaluations, in fact, are considered a promising method for estimating muscle activity and function in individuals with TMD and may be of great resource in assessing the response to possible therapy, such as also orthodontic therapy [14]. In fact, in the event that orthodontic treatment causes a condition of tension at the level of the oral and peri-oral musculature, elevated electromyographic values and altered mandibular kinematics are recorded [15]. This allows the clinician to change the method of treatment or even discontinue it for a more muscular wellness condition, especially if the patient in question already has temporomandibular disorders.

In recent years, there has been growing interest in the orthodontic field in the use of elastodontic devices. These are preformed silicone devices that can be used in both children and adults. Thanks to their structure, such devices are very easy to use, safe and comfortable for the patient. In fact, they allow stability to be achieved in the oral cavity with both esthetic and functional results. The activation of the device occurs with only chewing exercises on the part of the patient without the need for the orthodontist to resort to necessarily invasive orthodontic forces. On the other hand, there is an increasing demand for oral bites sold in pharmacies, which, since they do not require the collaboration of the orthodontist, are cost-effective. Many doubts arise, however, about their effectiveness at the level of oral health.

The aim of the study was to verify the clinical validity of two different gnatologic devices, the Eptamed elastodontic device and a common bite sold in pharmacies, by comparing the EMG and kinesiography activity of four pairs of masticatory and postural muscles (anterior temporalis, digastricus, masseters and sternocleidomastoids muscles) before and after 6 months of orthodontic treatment in adults patients with TMD. The clinical significance of that study would be to ensure that in the future, orthodontic treatments can be offered to patients with TMD with effective devices that do not worsen the health of the temporomandibular joint, which is already compromised, but rather provide benefits and muscle and joint relaxation.

2. Materials and Methods

This study was carried out in accordance with the fundamental principles of the Declaration of Helsinki and was approved by the Internal Review Board of the University of L'Aquila (Number 16137/2016). A total of 24 patients were enrolled in the study, and written informed consent was obtained from each subject. These subjects reported signs and symptoms of myofascial TMD according to group 1a of the Axis I of the Research Diagnostic Criteria for TMD (RDC/TMD) [16] and orthodontic treatment necessity for occlusal discrepancies according to the index of orthodontic treatment needs (IOTN) described by Brook and Shaw [17]. For each patient, a careful anamnesis and orthopanoramic exam were performed, intraoral and extraoral photos were taken and alginate impressions of the dental arches were made; finally, the dental technician was asked to cast the plaster models. The 24 patients were then included or excluded from the study according to the following inclusion and exclusion criteria.

The inclusion criteria were as follows:

(1) Age between 18 and 50 years;
(2) TMD diagnosis according to Axis I;
(3) Index of orthodontic treatment needs (IOTN) ≤ 3;
(4) Chronic orofacial pain (>3 months);
(5) Impairment sEMG activity after TENS according to Konchak et al. [18];
(6) Complete permanent dentition.

The exclusion criteria were as follows:

(1) Systemic diseases;
(2) History of local or general trauma;
(3) Neurological or psychiatric disorders;
(4) Pacemaker wearer;
(5) Episodes of epileptic seizures;
(6) Pregnancy;
(7) Assumed use of FANS, steroidal anti-inflammatory drugs, analgesics, SSRIs or opiates.

Based on the inclusion and exclusion criteria, 12 patients were eventually enrolled in the study, and the subjects were randomly divided into test and control groups through computer-generated software (https://www.sealedenvelope.com/ accessed on 12 March 2023). The control group comprised 6 patients (3 males, 3 females; aged less than 50 years) treated with a preformed bite, which is normally sold in pharmacies. The test group comprised 6 patients (3 males, 3 females; less than 50 years) treated with an Eptamed elastodontic device (EQ. UNIVERSAL). The two devices were very similar to each other and individually modified according to the patients' individual arches.

2.1. Instrumentation

Each patient in the study group received an EQ. UNIVERSAL device. This device is similar to a mouth guard, embracing both jaws, so it covers all dental elements up to the last molar in the arch. These devices are pre-formed removable silicone devices, so there are different sizes. To choose the device that best fits the size and shape of the dental arch, the distance between the palatal cusps of the first premolars is measured on the previously made plaster models. Patients were given the necessary instructions: the study group had to wear the device at night and only for 1 h and a half per day. During the day, patients had to perform chewing exercises to activate the device. The equilibrator is a type of orthodontic device that stimulates growth and, through the input of muscle movements, triggers tissue development towards proper chewing function. Biting this elastomeric device balances tension at the spheno-basilar synchondrosis, based on osteopathic medicine and philosophy [19]. Patients in the control group were given a preformed, self-molding resin splint normally available for purchase at low-cost pharmacies, matching the dental casts of both patients' arches obtained at the beginning of the study. As well as the case group, the control group was also asked to wear the device overnight and for an hour and a half a day with the same exercises to be performed.All examined patients underwent electromyographic examination and kinesiographic examination before the start of orthodontic treatment (T0) and after 6 months (T1); moreover, orthodontic examinations were performed for each patient periodically during the trial and thereafter. Electromyographic activity was recorded using an eight-channel Myotronics K7 Evaluation System (Seattle, WA, USA) and single-use electrodes (Duotrode, Ag-agcl surface bipolar electrodes, 20 mm center-interaxis, Myotronics-Noromed, Inc., Tukwila WA, USA). Electrodes were positioned on the left masseter (LMM), right masseter (RMM), left anterior temporal (LTA), right anterior temporal (LTA, RTA) [20], left digastric muscle (LDA), right digastric muscles (RDA) and the left and right sternocleidomastoid muscles (LSC, RSC) [21]. The sEMG and muscle activity records were expressed as the mean root square (RMS) of amplitude, expressed in μV and digitized with the K7 clinical software package (K7 Program 18.0, Myotronics-Noromed, Inc., Tukwila WA, USA, 2000) [2]. In this study, the authors evaluated scan 9 with muscle tone being evaluated with eyes closed and then with eyes open. This scan is very important because it highlights the muscle tone in rest position with closed eyes and after with opened eyes. In addition to scan 9, a kinesiographer was used to assess the kinematics of the mandibular in frontal, sagittal and lateral vision (K7/CMS®; Myotronics-Noromed, Inc., Tukwila, WA, USA) with an accuracy of 0.1 mm. An array containing 8 magnetic sensors tracked the motion of a 0.1 oz magnet (CMS Magnet; Myotronics-Noromed, Inc., Tukwila, WA, USA) that was attached to the labial gingiva beneath the mandibular incisor teeth with an adhesive gel. The kinesiographic examination was carried out by recording scan 1. It evaluates the degree of maximal voluntary mandibular opening, whereby the patient is asked to open and close his mouth to the maximum with eyes open. The degree of opening is expressed in millimeters. The data obtained from the electromyographic and kinesiographic examination were digitized and analyzed by means of software installed in a computer (K7 Program 18.0, Myotronics-Noromed, Inc., Tukwila, WA, USA, 2000).

2.2. Statistical Analysis

Electromyography microvolt values were recorded in an electronic spreadsheet. Subsequently, surface electromyography (sEMG) raw values from each muscle underwent a dual transformation through both averaging and summation, resulting in the creation of two new variables. Following this initial processing, each variable underwent a z-score transformation, calculated as the difference between the observed value and the mean of the sample, divided by the standard deviation of the sample. This normalization procedure was employed for the comparative evaluation of the data, effectively mitigating scale-related biases that may have existed between the averaged and summed values. To ascertain the statistical significance of the observed differences among groups, the study employed a non-parametric Wilcoxon signed-rank test. This statistical approach was used for its robustness in handling small sample sizes and non-normally distributed data. The predefined threshold for statistical significance was established at $p < 0.05$ for discerning meaningful distinctions in electromyographic activity between the groups under scrutiny.

3. Results

Baseline characteristics of the enrolled individuals showed no statistical difference in terms of age and sex, and all had a diagnosis of temporomandibular disorders according to Axis I of the RDC/TMD. This allows the two groups to be considered comparable.

The statistical analysis showed that at 6 months in closed-eye condition, there was an improvement in the electromyographic activity of all examined muscles. In particular, anterior temporalis, masseters, anterior digastricus and sternocleidomastoid muscles showed low values of microVolts in the Eptamed group compared with patients wearing standard bite. This finding is consistent on both sides (Figure 1). On the other hand in open-eye condition, in the Eptamed group compared with patients wearing standard bite, there was a statistically significant improvement in the electromyographic activity of only few muscles ($p < 0.05$): the left and right temporalis (LTA_OA, $p = 0.0022$; RTA_OA, $p = 0.0022$), the right and left digastric muscles (LDA_OA, $p = 0.014$; RDA_OA, $p = 0.0095$) and the left sternocleidomastoid (LTP_OA, $p = 0.005$) showed low values of microVolts. There was no significant improvement in masseter muscles and the right sternocleidomastoid muscle (Figure 1).

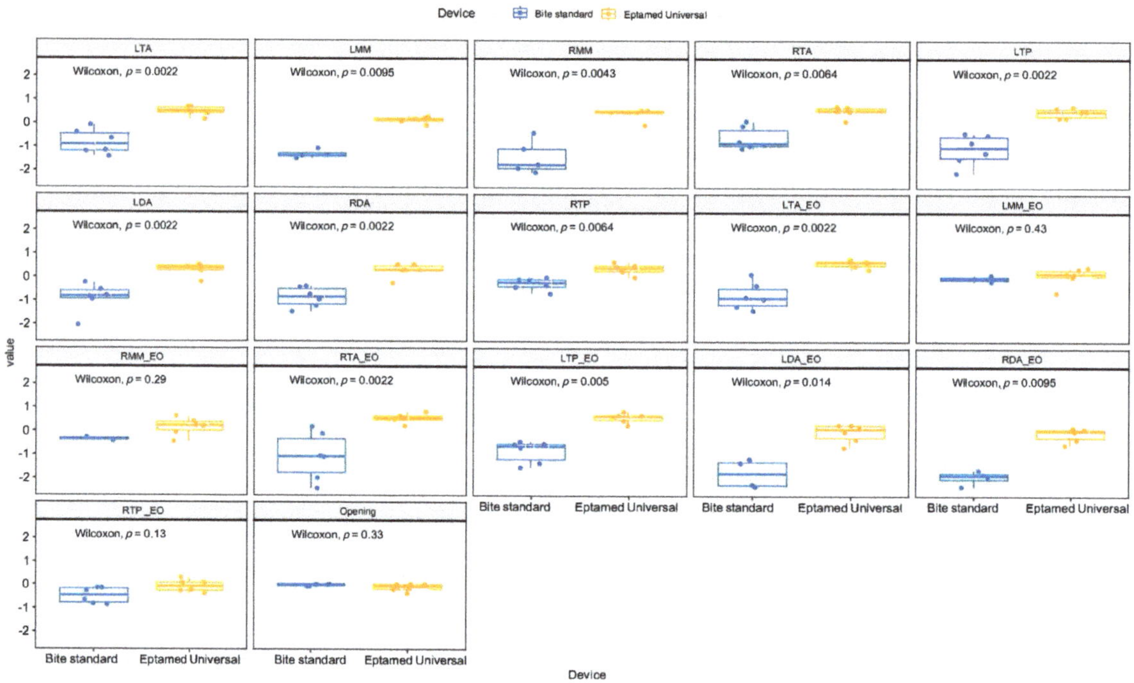

Figure 1. Non−parametric Wilcoxon signed-rank test performed following electromyographic examination in the open and closed eye condition of all muscles examined in the Eptamed group (in yellow) and the standard bite group (in blue) after 6 months of treatment.

Figure 2 describes the degree of maximum mouth opening, assessed by kinesiographic examination (scan 1) before the start of treatment (T0) and after 6 months of orthodontic treatment (T1) in the two groups. At T0, there was no statistically significant difference in the degree of maximum mouth opening in the two groups. After 6 months from the start of orthodontic treatment, in both Eptamed-treated and standard bite-treated patients, there was no change in the degree of mouth opening compared with T0. So, there is no statistically significant difference in the degree of opening between T0 and T1 and between the two groups.

Table 1 shows the z-score transformation of each variable. A negative value indicates a worsening of the muscle relaxation state, while a positive value indicates an improvement after 6 months of treatment. Note the lower tension and, thus, greater muscle relaxation in patients wearing Eptamed compared with patients wearing standard bite. In particular, in the eyes-closed condition, patients with Eptamed manifest a general improvement in all muscles examined; upon the opening of the eye (EO), the greatest relaxation is found at the level of the temporalis muscles on both sides, while modest improvements are seen in all other muscles examined. No change is seen in the degree of mouth opening. In the standard bite group, on the other hand, the state of tension worsens upon both eye opening and closing, with a markedly at eye opening.

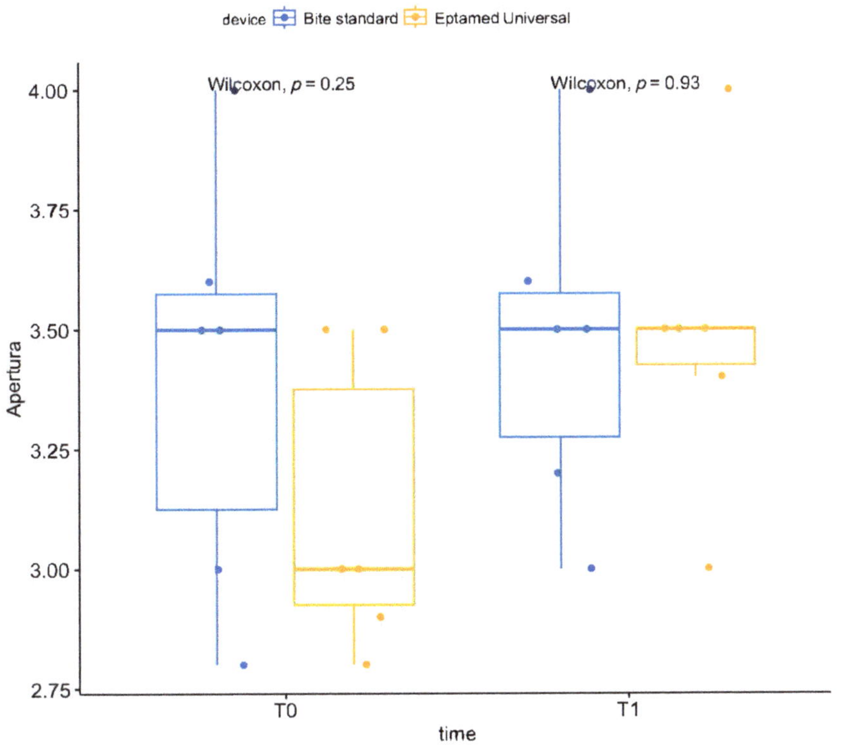

Figure 2. Non–parametric Wilcoxon signed-rank test performed following kinesiographic examinations in the condition of maximum mouth opening (scan 1) in the two groups.

Table 1. z−score transformation of each variable calculated as the difference between the observed value and the mean of the sample, divided by the standard deviation of the sample, described with a heatmap.

Id	Device	Lta	Lmm	Rmm	Rta	Ltp	Lda	Rda	Rtp	Lta_eo	Lmm_eo	Rmm_eo	Rta_eo	Ltp_eo	Lda_eo	Rda_eo	Rtp_eo	Opening
1	BITE	−1.21	−1.07	−1.75	−0.89	−0.40	−0.54	−0.95	−0.42	−0.91	−4.46	−4.85	−2.44	−0.53	−2.35	−3.71	−0.86	0.00
2	BITE	−1.17	−4.00	−3.30	−1.06	−0.46	−2.04	−1.46	−0.13	−1.23	−4.50	−3.30	−1.06	−0.44	−4.11	−3.09	−0.27	0.00
3	BITE	−0.10	−1.50	−0.40	0.11	−2.09	−0.85	−0.74	−0.69	0.14	0.11	−0.29	0.18	−1.54	−1.17	−1.64	−0.16	0.00
4	BITE	−1.42	−1.36	−2.08	−0.10	−0.79	−0.25	−0.43	0.00	−1.40	−5.00	−4.83	−2.00	−0.56	−1.33	−2.33	−0.67	−0.07
5	BITE	−0.67	−1.38	−1.08	−0.76	−1.23	−0.80	−0.39	−0.32	−0.33	−4.91	−3.55	−1.11	−0.71	−2.26	−1.92	−0.15	−0.07
6	BITE	−0.41	−3.00	−1.91	−0.94	−1.50	−0.97	−1.22	−0.09	−0.83	−0.15	−0.43	−0.11	−1.36	−3.50	−1.83	−0.82	0.00
1	EQ. U.	0.69	0.06	0.55	0.71	0.74	0.25	0.27	0.64	0.78	0.38	0.62	0.80	0.67	−0.33	−0.33	−0.28	−0.21
2	EQ. U.	0.14	−0.10	−0.08	0.11	0.26	−0.21	−0.27	0.00	0.34	−0.64	−0.47	0.20	0.24	0.00	0.03	−0.24	−0.17
3	EQ. U.	0.44	0.19	0.47	0.74	0.68	0.46	0.52	0.24	0.68	0.04	−0.07	0.64	0.83	−0.67	−0.56	−0.39	0.00
4	EQ. U.	0.66	0.15	0.55	0.54	0.53	0.42	0.53	0.80	0.70	0.45	0.24	0.58	0.67	0.29	0.07	0.07	0.00
5	EQ. U.	0.40	0.17	0.46	0.59	0.53	0.51	0.29	0.33	0.63	0.18	0.41	0.50	0.61	0.22	0.12	0.28	0.00
6	EQ. U.	0.55	0.27	0.48	0.52	0.25	0.32	0.30	0.44	0.49	0.19	0.20	0.46	0.44	0.29	0.17	0.05	−0.38

4. Discussion

From the data obtained, it can be concluded that

- The elastodontic device (EQ UNIVERSAL, Eptamed srl Via Ravennate, 979, 47522 Cesena (FC) ITALY) is capable of causing, in patients with temporomandibular disorders, a reduction in muscle tone at rest (scan 9) of the examined muscles (masseter, anterior temporal, digastric and sternocleidomastoid muscles) compared to a standard bite sold in pharmacies that, instead, causes an increase in muscle tension.
- The reduction of electromyographic activity with Eptamed is greater in the closed-eye condition than in the open-eye condition when scan 9 is recorded.
- No statistically significant difference was assessed during the kinesiographic examination at the maximum opening of the mouth in the comparison between the two groups at T0 and T1.
- No improvement in the electrical activity of the muscles was observed after 6 months of the use of the standard bite.

Following the results obtained from the above-mentioned study, the Eptamed device can be considered as a device to be administered to patients with TMD and orofacial pain before they start orthodontic therapy, as it is able to relax the stomatognathic muscles. These results are in line with the findings of Ortu et al. in which, comparing the electromyographic and kinesiographic examination of two elastodontic devices, it was seen that both ensured muscle relaxation, reducing the risk of TMD(s) development in treated subjects [22].

A few articles in the literature over the years called orthodontic treatment into question as a possible cause of TMD. Kononen et al. examined 166 children after orthodontic treatment, and 52% of them manifested signs and symptoms of TMDs [23]. Hirata et al. had shown that the development of dysfunction occurred equally in both orthodontically treated and untreated subjects, this means that orthodontic treatment does not always lead to the achievement of a muscular balance [24]. Also, Ricketts stated that clinical symptoms of joint derangement were noted as occlusions were changed, and he suggested that the various orthodontic forces provided during therapy may predispose patients to temporomandibular joint problems [25]. But, the most emblematic case is a 1987 Brimm vs. Malloy Michigan court case [26] in which an orthodontist was ordered to recover damages against a patient who had undergone mismanaged orthodontic treatment that caused her joint problems. Stohler and Zarb [27] state that orthodontic treatment of patients with TMD is considered radical and maximizes the risk of iatrogenic complications; that is, there is a significant risk of failure.

In this regard, guidelines based on the latest scientific evidence were drawn up, and an orthodontist must consider them in clinical practice [28]. In a nutshell, it is required of dentists to conduct a thorough examination and/or screening of TMD (presence of clicks, joint pain or mandibular deviations [29]) during orthodontic consultation and before starting orthodontic treatment both in patients who already have temporomandibular disorders and in healthy patients. However, it is necessary to stop any treatment if symptoms worsen (in patients diagnosed with TMD) or arise (in patients without TMD before treatment). In this way, the orthodontist can avoid both the worsening and the occurrence of TMD due to poorly managed orthodontic treatment.

In this circumstance, electromyographic and kinesiographic examination are essential tools that help the clinician assess whether a neuromuscular balance can be achieved following orthodontic treatment that does not compromise the health of the temporomandibular joint. According to Castroflorio et al., the electromyographic examination is considered an easily reproducible examination, so much so that the instrumentation used and a well-controlled sEMG protocol allow it to be used for longitudinal studies [30]. The presence of generalized muscle relaxation in all eight muscles analyzed by electromyographic examination is an indication of the effectiveness of the orthodontic treatment itself [22]. For this reason, the achievement of low muscle activity at rest must be a desirable outcome during treatment.

Kecik et al. [31] compared the dental and muscular changes that patients manifested following treatment of maxillary expansion with the quad-helix. In this case, in addition to clinical and cephalometric examinations, the need to contemplate electromyographic evaluations in the diagnostic setting was affirmed. The sEMG allowed appropriate diagnosis and prognosis to be formulated and also to monitor the functional impact of orthodontic therapies at various stages of treatment. In this way, the clinician can evaluate whether or not the orthodontic device is suitable in the presence of TMDs, thus avoiding worsening pain and dysfunction at the joint level.

To our knowledge, this is the first study to evaluate electromyographic and kinesiographic activity in subjects with TMD by comparing two different orthodontic devices, with the aim of assessing any muscle tension that these devices may cause. The use of the Eptamed device resulted in the achievement of greater muscle relaxation and a reduction in most of the electromyographic values considered. This did not occur in TMD patients wearing a bite sold in pharmacies. The improvement in electromyographic values in the case group is manifested mostly in the eyes-closed condition compared to the eyes-open condition. The change in electromyographic values according to visual input is in line with accumulated evidence indicating that TMD subjects show a dysregulation of the systems that control the response of the autonomic and somatomotor systems to sensory stimuli [32]. In a study by Monaco et al. [33], subjects with TMD in the open-eye condition had higher electromyographic values than healthy subjects in the same condition. These values increased by far even after standard spectacle wrapping. The lower improvement in values in the open-eye condition that occurred in our study, therefore, may be an intrinsic factor in the pathophysiology of temporomandibular disorders. TMDs are, in fact, classified as CSS and, as such, manifest a state of hyperexcitation and altered response to peripheral sensory stimuli, including those coming from the visual system [34].

However, the subjects who wore Eptamed at the opening of the eyes had electromyographic values worse than the closed eyes condition but better than the control group. This may suggest that elastodontic devices may, in this case, positively influence the responses of TMD patients to visual sensory inputs. On the other hand, the use of oral bites in orthodontic treatment in patients with TMD is controversial; several studies evaluated muscle electromyographic activity following the use of such devices, and the results are so variable that no firm conclusion can be made [35,36]. In fact, oral bites sold in pharmacies, by not requiring the cooperation of the dentist, lead the patient to a self-diagnosis that can be harmful. In fact, as was seen in this study, such devices may even worsen the patient's clinical situation, so they need careful evaluation and use.

It is important to note that this is a pilot study, for which it inevitably has some limitations: it must be considered that patients were evaluated only for 6 months of orthodontic treatment; it would be advisable to extend the follow-up to at least 1 year. In addition, few people were selected, so the sample is too small. Future studies should, therefore, overcome these limitations by increasing the sample size and observation time of the sample.

5. Conclusions

Our pilot study suggests that the Eptamed device could be a viable orthodontic device for use in individuals with TMD as it provides a reduction in muscle tension, which individuals with TMD can present, compared to a standard bite. However, due to the limited number of patients examined, further research will be needed in the future.

Author Contributions: Conceptualization E.O., S.C. and A.M.; methodology D.P. and S.C.; validation A.M.; writing—original draft preparation S.D.N. and E.O.; writing—review and editing S.D.N., L.D.S. and E.O. All authors have read and agreed to the published version of the manuscript.

Funding: This research received no external funding.

Institutional Review Board Statement: This study was carried out in accordance with the fundamental principles of the Declaration of Helsinki and was approved on 17 May 2016 by the Internal Review Board of the University of L'Aquila (Number 16137/2016).

Informed Consent Statement: Informed consent was obtained from all participants involved in the study.

Data Availability Statement: All the experimental data to support the findings of this study are available by contacting the corresponding author upon request.

Conflicts of Interest: The authors declare no conflicts of interest.

References

1. Ohrbach, R.; Bair, E.; Fillingim, R.B.; Gonzalez, Y.; Gordon, S.M.; Lim, P.F.; Ribeiro-Dasilva, M.; Diatchenko, L.; Dubner, R.; Greenspan, J.D.; et al. Clinical orofacial characteristics associated with risk of first-onset TMD: The OPPERA prospective cohort study. *J. Pain.* **2013**, *14* (Suppl. S12), T33–T50. [CrossRef]
2. Sergio Paduano, M.D.; Rosaria Bucci, D.D.; Roberto Rongo, D.D.; Silva, R.; Michelotti, A. Prevalence of temporomandibular disorders and oral parafunctions in adolescents from public schools in Southern Italy. *CRANIO®* **2018**, *38*, 370–375. [CrossRef]
3. Dworkin, S.F.; Von Korff, M.R.; LeResche, L. Multiple pains and psychiatric disturbance: An epidemiologic investigation. *Arch. Gen. Psychiat.* **1990**, *47*, 239–244. [CrossRef]
4. Manfredini, D.; Ahlberg, J.; Winocur, E.; Guarda-Nardini, L.; Lobbezoo, F. Correlation of RDC/TMD axis I diagnoses and axis II pain-related disability. A multicenter study. *Clin. Oral Investig.* **2011**, *15*, 749–756. [CrossRef]
5. Reis, P.H.F.; Laxe, L.A.C.; Lacerda-Santos, R.; Münchow, E.A. Distribution of anxiety and depression among different subtypes of temporomandibular disorder: A systematic review and meta-analysis. *J. Oral Rehabil.* **2022**, *49*, 754–767. [CrossRef] [PubMed]
6. Yap, A.U.; Chen, C.; Wong, H.C.; Yow, M.; Tan, E. Temporomandibular disorders in prospective orthodontic patients. *Angle Orthod.* **2021**, *91*, 377–383. [CrossRef] [PubMed]
7. de Leeuw, R.; Klasser, G. (Eds.) *Differential Diagnosis and Management of TMDs. Orofacial Pain: Guidelines, Assessment, Diagnosis, and Management*, 6th ed.; Quintessence Publishing: Berlin, Germany, 2018; pp. 144–207.
8. Andrews, L.F. *Straightwire: The Concept and Appliance*; LA Wells: San Diego, CA, USA, 1989.
9. Masci, C.; Ciarrocchi, I.; Spadaro, A.; Necozione, S.; Marci, M.C.; Monaco, A. Does orthodontic treatment provide a real functional improvement? A case control study. *BMC Oral Health* **2013**, *13*, 57. [CrossRef] [PubMed]
10. LeResche, L. Epidemiology of Temporomandibular disorders: Implication for investigation of etiologic factors. *Crit. Rev. Oral Biol. Med.* **1997**, *8*, 291–305. [CrossRef] [PubMed]
11. Popa, C.; Solomon, S.M.; Rudnic, I.; Mârţu, I.; Luchian, I.; Mârţu, M.A.; Sava, N.; Mârţu, S. Evaluation of oclusal trauma as a risk factor in the ethiology of chronic periodontitis. *Int. J. Med. Dent.* **2018**, *8*, 83.
12. Popa, C.G.; Luchian, I.; Ioanid, N.; Goriuc, A.; Martu, I.; Bosinceanu, D.; Martu, M.A.; Tirca, T.; Martu, S. ELISA Evaluation of RANKL Levels in Gingival Fluid in Patients with Periodontitis and Occlusal Trauma. *Rev. Chim.* **2018**, *69*, 1578–1580. [CrossRef]
13. Al-Saleh, M.A.; Flores-Mir, C.; Thie, N.M. Electromyography in diagnosing temporomandibular disorders. *J. Am. Dent. Assoc.* **2012**, *143*, 351–362. [CrossRef] [PubMed]
14. Santana-Mora, U.; López-Ratón, M.; Mora, M.J.; Cadarso-Suárez, C.; López-Cedrún, J.; Santana-Penín, U. Surface raw electromyography has a moderate discriminatory capacity for differentiating between healthy individuals and those with TMD: A diagnostic study. *J. Electromyogr. Kinesiol.* **2014**, *24*, 332–340. [CrossRef]
15. Giannini, L.; Maspero, C.; Batia, C.; Galbiati, G. Valutazione elettromiografica ed elettrognatografica del trattamento ortodontico-chirurgico. *Mondo Ortod.* **2011**, *36*, 12–28. [CrossRef]
16. Dworkin, S.F.; Le Resche, L. Research diagnostic criteria for temporomandibular disorders: Review, criteria, examinations and specifications, critique. *J. Craniomandib. Disord.* **1992**, *6*, 302–355.
17. Brook, P.H.; Shaw, W.C. The development of an index of orthodontic treatment priority. *Eur. J. Orthod.* **1989**, *11*, 309–320. [CrossRef]
18. Konchak, P.A.; Thomas, N.R.; Lanigan, D.T.; Devon, R.M. Freeway space measurement using mandibular kinesiograph and EMG before and after TENS. *Angle Orthod.* **1988**, *58*, 343–350. [PubMed]
19. Ortu, E.; Pietropaoli, D.; Cova, S.; Marci, M.C.; Monaco, A. Efficacy of elastodontic devices in overjet and overbite reduction assessed by computer-aid evaluation. *BMC Oral Health* **2021**, *21*, 269. [CrossRef]
20. Castroflorio, T.; Farina, D.; Bottin, A.; Piancino, M.G.; Bracco, P.; Merletti, R. Surface EMG of jaw elevator muscles: Effect of electrode location and inter-electrode distance. *J. Oral Rehabil.* **2005**, *32*, 411–417. [CrossRef]
21. Castro, H.A.; Resende, L.A.; Berzin, F.; Konig, B. Electromyographic analysis of the superior belly of the omohyoid muscle and anterior belly of the digastric muscle in tongue and head movements. *J. Electromyogr. Kinesiol.* **1999**, *9*, 229–232. [CrossRef]
22. Ortu, E.; Barrucci, G.; Aprile, G.; Guerrini, L.; Pietropaoli, D.; Monaco, A. Electromyographic evaluation during orthodontic therapy: Comparison of two elastodontic devices. *J. Biol. Regul. Homeost. Agents* **2020**, *34*, 1935–1939. [CrossRef]
23. Kononen, M.; Nystrom, M.; Kleemola-Kujola, E.; Kataja, M.; Evalahti, M.; Pekka, L.; Peck, L. Signs and symptoms of craniomandibular disorders in a series of Finnish children. *Acta Odontol. Scand.* **1987**, *45*, 109–114. [CrossRef] [PubMed]
24. Hirata, R.H.; Heft, M.W.; Hernandez, B.; King, G.J. Longitudinal study of signs of temporomandibular disorders (TMD) in orthodontically treated and nontreated groups. *Am. J. Orthod. Dentofacial. Orthop.* **1992**, *101*, 35–40. [CrossRef] [PubMed]
25. Ricketts, R.M. Clinical implications of the temporomandibular joint. *Am. J. Orthod.* **1966**, *52*, 416–439. [CrossRef] [PubMed]

26. Pollack, B. Cases of note: Michigan jury awards $850,000 in ortho case: A tempest in a teapot. *J. Mich. Dent. Assoc.* **1988**, *70*, 540–542. [PubMed]
27. Stohler, C.S.; Zarb, G.A. On the management of temporomandibular disorders: A plea for a low-tech, high-prudence therapeutic approach. *J. Orofac. Pain* **1999**, *13*, 255–261. [PubMed]
28. Sanjivan, K.; Donald, J.; Rinchuse, C.S.; Greene, L.; Johnston, E. Temporomandibular disorders and orthodontics: What have we learned from 1992–2022? *Am. J. Orthod. Dentofac. Orthop.* **2022**, *161*, 769–774. [CrossRef]
29. Machen, D.E. Legal aspects of orthodontic practice: Risk management concepts. Excellent diagnostic informed consent practice and record keeping make a difference. *Am. J. Orthod. Dentofacial. Orthop.* **1990**, *98*, 381–382. [CrossRef]
30. Castroflorio, T.; Titolo, C.; Deregibus, A.; Debernardi, C.; Bracco, P. The Orthodontic Treatment of TMD Patients: EMG Effects of a Functional Appliance. *CRANIO®* **2007**, *25*, 206–212. [CrossRef]
31. Kecik, D.; Kocadereli, I.; Saatci, I. Evaluation of the treatment changes of functional posterior crossbite in the mixed dentition. *Am. J. Orthod. Dentofac. Orthop.* **2007**, *131*, 202–215. [CrossRef]
32. Chen, H.; Nackley, A.; Miller, V.; Diatchenko, L.; Maixner, W. Multisystem dysregulation in painful temporomandibular disorders. *J. Pain* **2013**, *14*, 983–996. [CrossRef]
33. Monaco, A.; Ortu, E.; Giannoni, M.; D'Andrea, P.; Cattaneo, R.; Mummolo, A.; Pietropaoli, D. Standard Correction of Vision Worsens EMG Activity of Pericranial Muscles in Chronic TMD Subjects. *Pain Res Manag.* **2020**, *2020*, 3932476. [CrossRef] [PubMed]
34. Kindler, L.L.; Bennett, R.M.; Jones, K.D. Central sensitivity syndromes: Mounting pathophysiologic evidence to link fibromyalgia with other common chronic pain disorders. *Pain Manag. Nurs.* **2011**, *12*, 15–24. [CrossRef] [PubMed]
35. Tartaglia, G.M.; Lodetti, G.; Paiva, G.; De Felicio, C.M.; Sforza, C. Surface electromyograpic assessment of patients with long lasting temporomandibular joint disorders pain. *J. Electromyogr. Kinesiol.* **2011**, *21*, 659–664. [CrossRef]
36. Adibi, S.S.; Ogbureke, E.I.; Minavi, B.B.; Ogbureke, K.U. Why use oral splints for temporomandibular disorders(TMDs)? *Tex. Dent. J.* **2014**, *131*, 450–455. [PubMed]

applied sciences

MDPI

Article

Use of Tens in the Diagnosis of Functional Mandibular Lateral Deviation

Eleonora Ortu *, Sara Di Nicolantonio, Alessandra Mummolo, Ruggero Cattaneo, Davide Pietropaoli and Annalisa Monaco

Dental Unit, MESVA Department, University of L'Aquila, Piazzale Salvatore Tommasi 11, 67100 L'Aquila, Italy
* Correspondence: eleonora.ortu@univaq.it; Tel.: +39-0862434974

Abstract: Introduction: The traditional diagnosis of functional lateral deviation is based on a clinical evaluation and dental casts, supported by an instrumental analysis through X-rays; however, these diagnostic techniques do not provide any information about the neuromuscular system. Several years ago, some authors stressed the importance of the mandibular rest position and its reproducibility as a diagnostic reference, and this became possible with the development of the Myomonitor, TENS. Aim: The aim of this study was to compare mandible position before and after the use of ultra-low-frequency transcutaneous electric nervous stimulation (ULFTENS) in children with diagnosed functional mandibular lateral deviation. Methods: This study was performed on 60 children, aged between 8 and 13 years, with a mean age of 10.1 years (SD 0.81), and with functional mandibular lateral deviation diagnosed clinically, who were referred to the dental clinic for pediatric dental care. Diagnostic neuromuscular registrations were made for all children, and their casts were mounted on a Galetti articulator at the myocentric position. These casts were then compared to those provided by a wax bite registration in the intercuspidal position. Results: Compared with the existing intercuspidal position, neuromuscular registration showed improvement in 30 (50%) patients, 18 patients (30%) showed no changes, and worsening of the tooth midline discrepancy was assessed in the remaining 12 (20%). The molar relationship did not follow the same trend of the midline because of the three-dimensional changes in the maxillo–mandibular relationship induced by TENS. After TENS, there was a significant correlation between the midline and right side deviation ($r > 0.65$); there was no correlation between the midline and the left side ($r < 0.65$). Furthermore, the right molar movement showed no correlation with the contralateral molar ($r < 0.65$). The posterior areas of the arch moved in a very unpredictable way, resulting in the diagnosis and prognosis of mandibular lateral deviation as absolutely individual and unpredictable. Conclusions: This study suggests that TENS-recorded occlusion is an interesting diagnostic approach in orthodontics since it allows for visualizing the trends of the neuromuscular system.

Keywords: TENS; lateral deviation; myocentric position; growing patients

Citation: Ortu, E.; Di Nicolantonio, S.; Mummolo, A.; Cattaneo, R.; Pietropaoli, D.; Monaco, A. Use of Tens in the Diagnosis of Functional Mandibular Lateral Deviation. *Appl. Sci.* **2023**, *13*, 13258. https://doi.org/10.3390/app132413258

Academic Editors: Giuseppe Minervini, Conrado Aparicio and Rocco Franco

Received: 24 October 2023
Revised: 28 November 2023
Accepted: 8 December 2023
Published: 14 December 2023

1. Introduction

Mandibular laterodeviation is a condition in which the mandible and maxilla of a patient are not perfectly aligned, presenting a lateral deviation. There are two main methods for treating mandibular laterodeviation: non-surgical treatment (fixed and removable orthodontic appliances) and surgical treatment (orthopedic surgery and orthognathic surgery) [1]. On the basis of the etiology, they can be divided into structural or functional, and the latter is distinguished from the former both because they recognize different causes of origin and for the lack of lesions or structural alterations; the jaw, in case of precontact, to find the occlusal boarding, rising from the resting position, moves laterally to avoid the occlusal interference by establishing an avoidance reflex and modifying its starting position [2]. Functional lateral deviation can be the effect of a cross or a scissor bite due to the reduction of the transverse diameter of the palate, but could also be due to a neuromuscular asymmetry because of birth trauma associated with muscular torticollis and/or ocular defects [3–7]. The traditional diagnosis of functional lateral deviation is based on a clinical evaluation and dental casts, supported by an instrumental analysis through X-rays in the postero–anterior projection; however, these diagnostic techniques do not provide any information about the neuromuscular system [8,9].

Among the parameters used for orthodontic diagnostic evaluation, the vertical dimension of occlusion, the antero–posterior and latero–lateral reciprocal relationship of the dental arches are of fundamental clinical importance. Variations in the spatial relationship between the upper and lower arches, compared to ideal standards, are relevant factors in the design of the orthodontic treatment plan [10]

Typically, orthodontic treatment of a functional lateral deviation consists of therapy with different devices that align the jaw.

Often, orthodontic treatment, despite possible immediate aesthetic success, determines a constrained occlusal position that could determine long-term stability problems; occlusion does not require neuromuscular adaptation, but rather, orthodontic treatment must adapt to the neuromuscular system and not vice versa [11,12].

One of the possible hypotheses regarding the instability of orthodontic results could be that the neuromuscular balance was not taken into account during the diagnosis and treatment of laterodeviations. Thus, orthodontic treatment could achieve an aesthetic result at the expense of neuromuscular balance. This could lead to instability of the result in the medium and long term. In general terms, it could be suggested that orthodontic therapy should achieve an occlusion that does not require neuromuscular adaptation and that the therapy itself should adapt to the neuromuscular system.

Several years ago, some authors stressed the importance of the mandibular rest position and its reproducibility as a diagnostic reference [13,14].

According to this idea, the mandibular rest position is largely determined by the spatial position in which the occlusal contact occurs. Therefore, the mandibular rest position is usually "adapted" by the neuromuscular system, which will keep the mandible in a position appropriate for stomatognathic functions. A recent literature review indicates that there is strong clinical evidence that Rest Mandible Position (RMP), Rest Vertical Dimension (RVD), and the Interocclusal Dimension (IOD) are to be found within a three-dimensional spatial "range" largely determined by the active adaptation of the neuromuscular system [15].

This adaptation is extensive, but in specific cases, probably associated with other not necessarily occlusal components, dimensional variations may go beyond the limits of muscular compensation and induce occlusal instability and/or acute and chronic suffering [16,17].

The basic idea of these observations is that starting from a relaxed neuromuscular resting position of the mandible, it is possible, by comparing it with the habitual occlusion, to assess the three-dimensional occlusal space in which the contact between the jaws should take place without an adaptation effort of the stomatognathic musculature.

The use of dental ULFTENS is among the techniques used to achieve muscle relaxation and a resting position of the jaw. Dental ULFTENS belongs to the category of TENS that involves, in a general sense, the cutaneous stimulation of nerve branches located more proximally to the electrode application site. The mechanism of action of such stimulation has been extensively studied and demonstrated and involves, depending on the frequency of delivery, the intervention of different response modulation centers. Stimulation with frequencies below 20 Hz is referred to as low-frequency TENS, and below 4 Hz is referred to as ultra-low-frequency TENS, also called "acupuncture-like" in some cases [18,19]. In the case of low and ultra-low-frequency TENS, a central mechanism of action mediated by the prefrontal cortex and the ventrolateral PAG has been suggested. This action involves, among others, an endorphinic activity at the PAG level, a peripheral response characterized by modulation of central pain pathways, and a reduction in muscle tone together with a lowering of sympathetic tone [20]. Transcutaneous electrical nerve stimulation and conditioned pain modulation influence the perception of pain in humans [21].

Dental ULFTENS consists of transcutaneous electrical stimulation with an electrostimulation of the V and VII pairs of cranial nerves. It is performed through the application of electrodes bilaterally at the level of the area anterior to the tragus. The stimulation frequency is of the order of 0.66, which is why it belongs to the ultra-low category, and the amplitude of this stimulation is such that it evokes a minimal passive upward movement of the jaw. This is determined by the direct action of the impulse that reaches deep into the mandibular branch of the trigeminal, through the coronoid incisura, exciting its motor fibers and causing peripherally in a dromic way a contraction of the elevator muscles of the mandible. In this way, it would act according to a central mechanism as mentioned above and associate an automatic peripheral muscular upward movement of the mandible. Dental ULFTENS has been shown to be able to reduce the myoelectric activity of the masseter and anterior temporalis muscles in a resting position [22]. This technique of muscle stimulation and relaxation has already been used to evaluate sagittal relationships in an orthodontic population characterized as Class II division I with mandible dentoalveolar retrusion and allows for visualizing an unusual trend of growth. The work suggested that different orthodontic diagnoses can be hypothesized through ULFTENS depending on the individual subject's characteristic neuromuscular vector [1], in particular, through sensory amplitude stimulation, which is less uncomfortable for young patients [23]. It is therefore possible to hypothesize treatments aimed at the neuromuscular structure of the individual independently of the classical malocclusion pattern based on clinical, model, and radiogram observations. At this present time, there are no observations regarding the effects on the lateral plane of acutely administered ULFTENS in orthodontically aged subjects characterized by functional laterodeviation. It is conceivable that, as with the effects of ULFTENS in the sagittal plane, responses can be observed that are guided by the neuromuscular vector of the individual subject, irrespective of the type or direction of laterodeviation classified statically by clinical observation, plaster models, and radiograms.

The primary objective of TENS is to decondition and relax the muscles of the jaw and face in order to identify and establish the mandibular rest position with stomatognathic muscles relaxed. The aim of this study is the evaluation of changes in the occlusal position of the jaw after the relaxation procedure operated through the use of ULFTENS (ultra-low-frequency transcutaneous electric nervous stimulation) in children with diagnosed functional mandibular lateral deviation.

2. Material and Methods

This study was carried out according to the principles of the Helsinki Declaration. The University Ethics Committee approved the following project (16137/2016). Initially, in this study, 130 subjects between the ages of 8 and 13 were visited by the same operator at the dental clinic of the University of L'Aquila.

The inclusion criteria were as follows:

- Deviated chin from midsagittal plane (considering a plane drawn perpendicular to the bipupillary line and the commissural line);
- Misalignment between the upper and lower frenulum;
- Asymmetries between the upper and lower interincisal lines and between the molar and canine classes;
- Deviation of the median line in maximum intercuspidation, and in a resting position;
- Deviation of the jaw during mouth opening;
- Noises and "tenderness" of the temporomandibular joint,
- No orthodontic treatment performed prior to the visit;
- Deviation of the midline of the anterior teeth > 1.5 mm with open mouth alignment.

The exclusion criteria were as follows:

- Absence of written informed consent from parents/legal guardians;
- Presence of systemic diseases;
- Presence of epilepsy and/or pacemaker;
- Presence of skeletal asymmetry;

- Presence of posterior or lateral cross bite;
- Absence of congenital torticollis at birth;
- Presence of metabolic problems, or systemic or infectious diseases of the mother in the gestational period and drug or medication intake;
- Absence of dystocic delivery or preceded by the use of vasosuprine or oxytocin, and absence of traumatic practices during delivery (e.g., forceps) or breech delivery;
- Absence of major trauma to the face, skull, and neck during childhood that required medical intervention.

In the end, 60 patients were enrolled in this study (mean age 10.1 $\pm$ 0.81), and the protocol used was the one described below.

2.1. Experimental Protocol

Neuromuscular recording was performed on 60 patients. First, alginate impressions of the upper and lower arches were taken and accurate dental casts were made for each patient. The midline asymmetry was measured in relation to the upper and lower incisal midline. Secondly, the discrepancy of the midline with respect to the upper and lower interincisal lines was evaluated. For each plaster model then, reference points were highlighted: a line that goes from the interincisal point of the maxilla up to the mandible, and a line on the right and left molar side on the horizontal plane.

Later TENS was applied to relax the musculature and thus determine the rest position of the mandible, which was then measured and recorded.

TENS was administered using an electro-stimulator specifically designed for dentistry called Myomonitor J5 (J5, Myotronics, Kent, WA, USA, https://www.myotronics.com/j5, accessed on 4 December 2023).

The electrical pulse delivered has a frequency of 0.66 Hz, which is why this type of stimulation is also called ULF (ultra-low frequency) TENS. The duration of the impulse is less than 500 µs, and the amplitude is such that it does not evoke the upward movement of the mandible during the preparation phase (5 min) prior to recording the cranial jaw relationship. This amplitude of stimulation, defined as sensory amplitude, has been shown to be sufficient in inducing a reduction in the surface emg values of the anterior temporal and masseter muscles comparable to that obtained with motor amplitude ULFTENS, which requires a greater amplitude of stimulation and, therefore, more discomfort for the recipient. Stimulation is transferred to the subject by means of three electrodes, one positioned posteriorly on the neck (below the hairline, on the midline) and the other two anterior to the tragus, one on the right and the other on the left. The electrode on the back of the neck is the positive end of the stimulating circuit, while the negative end is split into the two anterior electrodes to allow symmetrical and simultaneous stimulation (Figure 1).

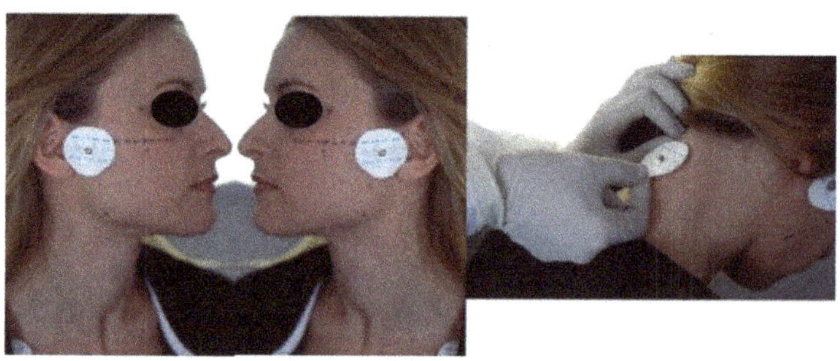

Figure 1. These photos show the application of TENS electrodes.

The stimulation protocol involves 5 min of sensory amplitude stimulation, then the person is asked to open the mouth and a previously prepared occlusal registration material is injected onto the occlusal surfaces with a disposable syringe without needle. In our case, this is a dental resin (Sapphire™ Acrylic Peripheral Impression Material, Keystone Industries, Gibbstown, NJ, USA, https://dental.keystoneindustries.com/product/sapphire-acrylic-peripheral-impression-material/, accessed on 7 December 2023).

The patient is asked to close the mouth and gently pull the lips together without manipulating the jaw or the face. One waits a minute for the jaw to return to the rest position obtained with ULFTENS and increases the amplitude of the impulse until a slight automatic upward movement of the jaw towards the upper jaw is obtained. This movement determines the impression of the teeth of the upper jaw in the resin placed on the occlusal surface. When the resin has a plastic consistency (approx. 1 min) it is removed from the mouth and placed on the plaster models awaiting its complete hardening.

At this point, the child is discharged. When hardening is complete, the models are mounted on the Galetti articulator(Sevtis-Fisat s.r.l. Plastic Technology, Tianjin, China). In Figure 2, there is shown a patient who does not change her position after TENS [1,24].

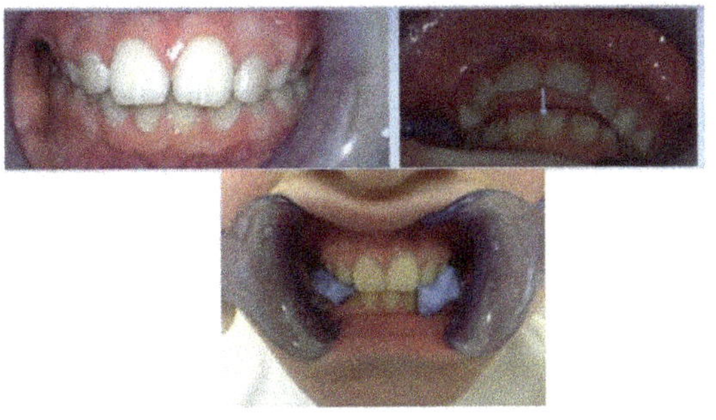

Figure 2. This patient does not change her jaw position after TENS stimulation, mandibular asymmetry on the right (blue arrow) is maintained after TENS.

With dental casts mounted in the articulator in the myocentric position with the interposed print, the change from the occlusal position was evaluated considering the initial reference lines; it was then calculated in millimeters and analyzed by a digital caliper. All measurements were performed twice, and the examiner's error was examined using the formula $s(i) = \sqrt{\Sigma d^2/2n}$. The intra observer methodological error $s(i)$ was calculated from duplicated recordings. S denotes the variance of the total sample of 60.

2.2. Statistical Analysis

The Stata package was used to perform a paired t-test for independent samples comparing mean and variance shifts of the midline, right molar side, and left molar side. Differences with a value of $p < 0.05$ were regarded as significant. A correlative tendency was analyzed with Pearson's correlation coefficient (r). A value of r equal to or greater than 0.65 was considered a positive correlation. Two sample proportion tests were carried out on movement direction comparing left or right shifts on the midline, with those on the right and left molar sides. Differences with a value of $p < 0.05$ were regarded as positively correlated. One sample proportion test was performed on each measurement (midline, right molar side, left molar side) considering 0.5 (50%) as the probability of left or right shift.

3. Results

After the TENS procedure, the midline shift (MMT) compared to the mandibular midline in habitual occlusion (MMHO) was 1.41 ± 1.24 on the right molar side (RMT), the shift was 1.76 ± 1.39, and the left molar side (LMT) was 1.10 ± 0.99 (Figure 3). No statistical significance was found comparing the midline mean (MMT) shift to the right molar side shift (RMT) (p = 0.13) and to the left molar side shift (LMT) (p = 0.24). A significant difference was revealed comparing the right molar side to the left molar side shift (p = 0.041). All conditions except one showed a low correlation (r < 0.65). Only the midline (MMT) and the right molar side amount of shift after TENS (RMT) showed a positive correlation (r < 0.65). The mandibular midline (in habitual occlusion) asymmetry is not significantly located on the left side compared to the right one ($p = 0.6$) (Figure 4). After the TENS procedure, 24 individuals shifted to the right of their midline, 21 to the left, and 15 did not change position. One sample proportion test failed to find a positive correlation between right and left shifts ($p = 0.29$) with TENS. The right molar side after TENS shifted distally in 36 individuals, mesially in 12, and it did not change in the remaining 12. A statistically significant difference was shown between the right and left side shifts on the molar side ($p = 0.018$). The left molar side after TENS shifted mesially in 21 individuals, distally in 18, and it did not change in 21. No statistically significant difference was found in the left molar side shift ($p = 0.39$) (Figure 5). After the TENS procedure, 30 subjects reduced midline asymmetry, 12 worsened, and 18 did not change midline asymmetry. Considering those patients whose midline asymmetry worsened or did not change and comparing them to those individuals with reduced asymmetry, the assessment of the intermaxillary relationship in terms of discrepancy from the midline did not show a statistically significant reduction of asymmetry after TENS ($p > 0.05$).

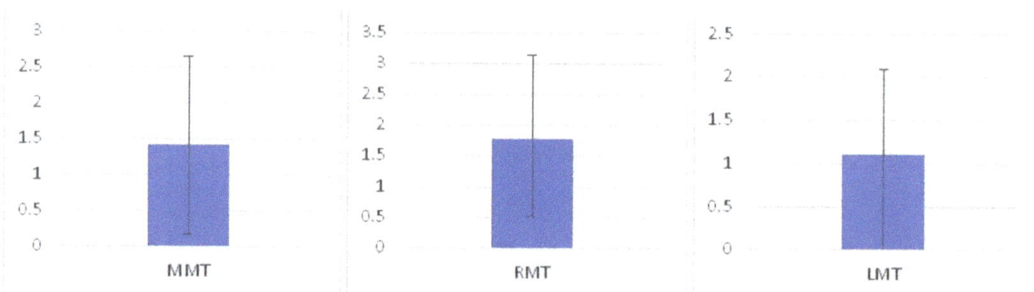

Figure 3. Mean value, standard deviation of midline mandibular position and right and left molar position after TENS.

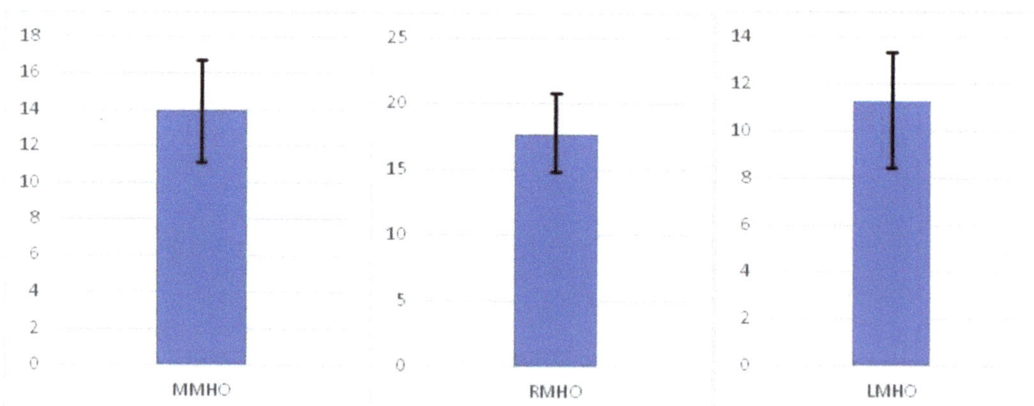

Figure 4. Mean value, standard deviation and paired comparison of midline mandibular in habitual occlusion (MMHO), right molar in habitual occlusion (RMHO), and left molar in habitual occlusion (LMHO).

The midline had the same shift direction compared to the left molar side in six individuals and the opposite shift direction in fifty-four individuals. The midline has the same shift direction compared to the right molar side in 12 individuals and the opposite shift direction in 48 individuals. The right–left molar side concordance was 30 out of 60 subjects; 30 out of 60 patients had the opposite shift direction on the left and right molar sides.

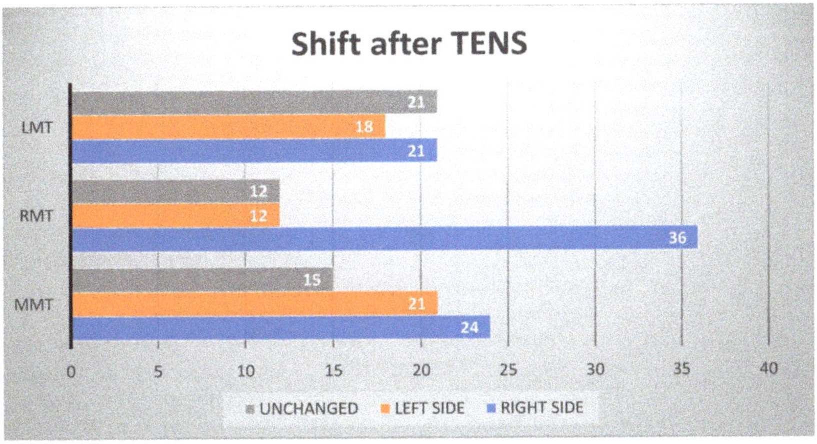

Figure 5. Number of shifts after TENS for subjects.

4. Discussion

Our results suggest that after muscle relaxation with dental sensory ULFTENS, the position of the jaw may or may not conform with the observation of laterodeviation made clinically or through plaster models through the position of the midline with closed teeth in habitual occlusion. Notably, in our study sample, a substantial group of the subjects worsens the midline asymmetry present in habitual occlusion, while another part does not change the asymmetry. Furthermore, considering the three-dimensional spatial displacement of the mandible as a whole, which can be represented by the simultaneous analysis of the position of the right and left molars, it can be observed that ULFTENS induces a resting position in which the right and left displacement in the laterolateral plane is frequently rather complex. For instance, the right side moves mesially, but the left end does not change position or it moves in a non-proportional way. If the mandible had simple translation movements on the horizontal plane, like those made on articulators or obtained with manual techniques on patients, the amount of lateral shift on the midline, left molar side, and right molar side would be proportional, and no significant difference in statistics would be found among the midline and left and right molar sides. This result suggests that functional lateral deviation could be an inhomogeneous diagnostic group and suggests that an approach with dental sensory ULFTENS to muscle balance could be useful in order to organize an actual functional orthodontic treatment. One of the assumptions of functional laterodeviations, in fact, is that in these subjects there exists a tonic imbalance of the musculature of the stomatognathic apparatus, and this would involve a deviated growth in relation to the continuous thrusts that the musculature would exert on the bone and dental matrix [25–28]. Our data with ULFTENS seem to confirm these observations. In fact, a substantial part of our sample shows that the muscular "relaxation" obtained with ULFTENS results in a different latero–lateral position of the mandible, a sign that, under "habitual" conditions, the "postural" and "functional" position of the mandible is maintained with an asymmetry of muscle tone. The muscle relaxation achieved with ULFTENS does not necessarily lead to a realignment of the midline and shows that the spatial relationships between the two arches can be rather complex, as already shown for the vertical and sagittal plane in a previous work [1]. In relation to what has been observed, it could be useful, in latero-functional deviations, to evaluate the resting position of the mandible with relaxed musculature and, therefore, not "conditioned" by the need to adapt to the skeletal and dental structure, especially when guided by the habitual occlusion generally used for orthodontic diagnosis. In this way, it may be possible to integrate the treatment plan and suggest a therapy more aimed at obtaining spatial planes and occlusal and structural relationships in tune with muscular symmetry. Dental TENS is a valid aid used to achieve muscle relaxation, and this relaxation has been described in electromyographic and kinesiographic terms such as reduction of myoelectric tone, variation of interocclusal free space, and occlusal position after TENS [29]. The use of this technique has the advantage that the resting position and the respective cranial–jaw relationship are obtained without direct manipulation of the jaw by an operator and without the direct will or voluntary mobility of the subject. The mechanism of action of TENS is complex and, although also present among its effects is that related to muscle tone on the musculature purely innervated by the V and VII pair of cranial nerves, some physiopathological responses to dental TENS may concern the overall capacity of adaptation to the environmental demands of the stimulated subject [30]. The displacement of the rest position after TENS is not predictable based on the usual dental and skeletal characteristics, and the response to stimulation appears to be individual, probably associated with the complex mechanisms linked to the growth and development of the individual. What has been observed allows us to suggest that frequently, in the functional lateral deviations, the usual occlusal position and rest position are a form of compensation that can be shown with the TENS stimulation.

The causes of the asymmetry of the craniofacial district and consequently also of the cranial mandibular district can be many and different. Some of them have a pre- or neonatal origin, others are post-birth, some have been correlated with functional or parafunctional activities of the stomatognathic system, and still others with adaptations of the stomatognathic system to extrastomatognathic and postural requests in general [31–36]. All these causes can contribute to the structural asymmetry that arose at the time of the action of the case.

What could be interesting to evaluate at a clinical level is the persistence of an "unfavorable" adaptation in terms of lateral deviation in that this adaptation in place could prelude to a clinical, aesthetic deterioration of lateral deviation or even prevent an orthodontic treatment traditional. In other words, a clinical/therapeutic classification of functional laterodeviations could find support in the ability to observe the "relaxability" of the musculature and the degree and direction of the maxillo–mandibular growth compensation vectors consequent to it. In this sense, clinicians have often stressed the need to take into consideration not only the skeletal aspect and the static relationship between the maxilla and the jaw, but also the fascial, muscular, and articular aspects and the functions of other districts of the body adjacent to and far from the stomatognathic district [37–40]. In this perspective, one of the main goals of orthodontic treatment is not only to achieve satisfactory aesthetic results but also to create a functional occlusion, meaning an occlusion that does not require a neuromuscular compensation of adaptation passing from the resting position to the occlusal one and does not overload the systems mentioned above [41–43]. Indeed, our work seems to suggest that the discrepancy between the resting position and habitual occlusion with respect to that obtainable with TENS is often in progress in children with functional lateral deviation and that, if it is shown that the musculature and the neuro fascial system after TENS is relaxed, it is necessary to work in relation to this neuro–muscular force and not contrary to it. In this way, the direction of the variation obtained with TENS should be taken into account in the treatment plan. It has been suggested that TENS acts on various physiological systems, including muscle, vision and accommodation, the autonomous one, pain control, and endorphins, and that, consequently, for this complex mechanism, its central action could be useful in the diagnosis and therapy of TM disorders [44]. This mechanism of action and its clinical effect could also be implemented in children characterized by functional lateral deviation. These observations suggest that the use of TENS could improve the diagnostic and therapeutic possibilities in orthodontics by indicating the neuromuscular state of the resting position of the mandible and its adaptation to habitual occlusion. This work has its limitations, among them being the lack of a classical diagnosis based on Angle's criteria. However, we must underline the fact that, within the logic in which it can be inscribed, the relevant aspect of our research was the functional lateral deviation that is independent of the type of dental class and skeletal class.

Another limitation of the work is that no comparison was made with other relaxation techniques of the stomatognathic musculature. On the other hand, our intention was to analyze the behavior of the resting position of the jaw precisely using the dental ULFTENS. The comparison with other techniques was not among the current aims of the work, although such a comparison could be the subject of new work. The recording of the cranio–mandibular relationship was obtained several times per child. This could result in different recordings leading to different resting and occlusal positions under TENS. In our study, the reproducibility of the recording was not performed through technical algorithms, but simply by observing the spatial "pattern" obtained. The type of displacement was always similar, so we chose to use the first recording in which the child is least "fatigued". On the other hand, this limitation is difficult to overcome when, even with other techniques, one tries to propose an occlusal dimension different from the usual one. In our work, the starting point is to obtain a resting position with relaxed musculature. In this work, we have tried to represent the clinical work as closely as possible. Performing several tests makes the technique time-consuming and investigative and does not ensure the proper cooperation of the young patients. This work, therefore, attempts to represent a research development of clinical practice. As a matter of fact, we are currently conducting a study to assess the reliability and reproducibility of the technique, and the preliminary data as a whole seem to confirm the indications that have emerged from this work. A further limitation of this work is due to the lack of electromyographic and kinesiographic control of the application of TENS. Although previous works confirm that EMG values are reduced on average and interocclusal free space increases after TENS, indicating a relaxation of the stomatognathic musculature, and the parameters of the autonomous system of the accommodation move towards a lesser sympathetic control, this work has not instrumentally recorded the effect of TENS on muscles, free space, and autonomous control [45]. On the other hand, the average values demonstrated elsewhere indicate that the TENS effect is generalizable and that only a minor sample can have a "paradoxical" effect, such as that reported by Konchak. In this sense, our work must be implemented with the use of other objective parameters of neuromuscular activity. Such research is difficult to perform with children due to its length and execution protocol and may not be decisive for the purposes of this research. The fact that the selected children did not present particular clinical disorders of the neuromuscular system, coordination, sight, and systems related to the stomatognathic system and that the research was concentrated on functional lateral deviations makes the possibility of misunderstanding the results less significant. Subsequent work will focus on the differences in response to the TENS of subjects with lateral deviations without the involvement of other districts (sight, posture, autonomous system, etc.) and subjects with definable "syndromic" forms in which other signs or symptoms of extrastomatognathic involvement are present.

5. Conclusions

Therefore, compared to the study conducted, the mandibular laterodeviations characterized by various degrees of the median line asymmetries and, above all, with asymmetric dynamics of the jaw must be evaluated and diagnosed considering both stomatognathic and extrastomatognathic aspects. In this regard, a protocol that considers the general causes of asymmetry through the anamnesis, the clinical examination, and the EMG/KNG evaluation activated with TENS makes it possible to diagnose the nature of the symmetry and the role of the dentist in the possibility of treating it.

Author Contributions: Methodology, E.O.; Investigation, A.M. (Annalisa Monaco), R.C. and A.M. (Alessandra Mummolo); Data curation, D.P.; Writing—review and editing, S.D.N. All authors have read and agreed to the published version of the manuscript.

Funding: This research received no external funding.

Institutional Review Board Statement: Ethics approval (number 16137/2016) was obtained by the IRB of the University of L'Aquila. The study protocol was drawn following the European Union Good Practice Rules and the Helsinki Declaration.

Informed Consent Statement: Written informed consent has been obtained from the patient(s) to publish this paper.

Data Availability Statement: The data presented in this study are available on request from the corresponding author. The data are not publicly available due to privacy.

Acknowledgments: The authors would like to thank Giovanna Ciampa and Gianmario Palmieri for their valuable collaboration in this research in data collection.

Conflicts of Interest: The authors declare no conflict of interest.

References

1. Monaco, A.; Cattaneo, R.; Spadaro, A.; Marzo, G. Neuromuscular diagnosis in orthodontics: Effects of TENS on the sagittal maxillo-mandibular relationship. *Eur. J. Paediatr. Dent.* **2008**, *9*, 163–169. [PubMed]
2. Cimic, S.; Zaja, M.; Kraljevic, S.; Simunkovic, M.; Kopic, A.; Catic, A. Influence of Occlusal Interference on the Mandibular Condylar Position. *Acta Stomatol. Croat.* **2016**, *50*, 116–121. [CrossRef] [PubMed]
3. Pirttiniemi, P.; Gron, M.; Alvesalo, L.; Heikkinen, T.; Osborne, R. Relationship of difficult forceps delivery to dental arches and occlusion. *Pediatr. Dent.* **1994**, *16*, 289–293. [PubMed]
4. Monaco, A.; Streni, O.; Marci, M.C.; Sabetti, L.; Marzo, G.; Giannoni, M. Relationship between mandibular deviation and ocular convergence. *J. Clin. Pediatr. Dent.* **2004**, *28*, 135–138. [CrossRef] [PubMed]
5. Kwak, Y.Y.; Jang, I.; Choi, D.S.; Cha, B.K. Functional evaluation of orthopedic and orthodontic treatment in a patient with unilateral posterior crossbite and facial asymmetry. *Korean J. Orthod.* **2014**, *44*, 143–153. [CrossRef] [PubMed]
6. Cardinal, L.; Martins, I.; Gribel, B.F.; Dominguez, G.C. Is there an asymmetry of the condylar and coronoid processes of the mandible in individuals with unilateral crossbite? *Angle Orthod.* **2018**, *89*, 464–469. [CrossRef] [PubMed]
7. Solani, B.; Talebian Ardestani, M.; Boroumand, H.; Ostadmohammadi, V.; Hallajnejad, M.; Kashani Zade, M.; Loghman, A.; Talebian Ardestani, A. Risk factors associated with positional plagiocephaly in healthy Iranian infants: A case-control study. *Iran. J. Child. Neurol.* **2022**, *16*, 85–92. [CrossRef] [PubMed]
8. Severt, T.R.; Proffit, W.R. The prevalence of facial asymmetry in the dentofacial deformities population at the University of North Carolina. *Int. J. Adult Orthod. Orthognath. Surg.* **1997**, *12*, 171–176.
9. Baek, S.H.; Cho, I.S.; Chang, Y.I.; Kim, M.J. Skeletodental factors affecting chin point deviation in female patients with class III malocclusion and facial asymmetry: A three-dimensional analysis using computed tomography. *Oral. Surg. Oral Med. Oral Pathol. Oral Radiol. Endod.* **2007**, *104*, 628–639. [CrossRef]
10. Osawa, K.; Nihara, J.; Nishiyama, H.; Takahashi, K.; Honda, A.; Atarashi, C.; Takagi, R.; Kobayashi, T.; Saito, I. A three-dimensional investigation of mandibular deviation in patients with mandibular prognathism. *Maxillofac. Plast. Reconstr. Surg.* **2023**, *45*, 4. [CrossRef]
11. Jankelson, R. *Neuromuscular Dental Diagnosis and Treatment*; Ishiyaku EuroAmerica: St. Louis, MI, USA, 2005; Volume II.
12. Manni, A.; Brunori, P.; Ursini, R.; Deli, R. Neuromuscular occlusion in the orthodontic treatment of craniomandibular disorders. A clinical case. *Minerva Stomatol.* **2002**, *51*, 95–101.
13. Steenks, M.H. Reference positions of the mandible. *Ned. Tijdschr. Tandheelkd.* **2009**, *116*, 376–381. [PubMed]
14. Lerman, M.D. A revised view of the dynamics, physiology, and treatment of occlusion: A new paradigm. *Cranio* **2004**, *22*, 50–63. [CrossRef] [PubMed]
15. Goldstein, G.; Goodacre, C.; MacGregor, K. Occlusal Vertical Dimension: Best Evidence Consensus Statement. *J. Prosthodont.* **2021**, *30*, 12–19. [CrossRef] [PubMed]
16. Calamita, M.; Coachman, C.; Sesma, N.; Kois, J. Occlusal vertical dimension: Treatment planning decisions and management considerations. *Int. J. Esthet. Dent.* **2019**, *14*, 166–181.
17. Constantinescu, F.E.; Savastano, F.; Perlea, P.; Constantinescu, M.V. Complete morphofunctional oral rehabilitation by physiological increase of occlusal vertical dimension according to computerized mandibular scanner. *Rom. J. Morphol. Embryol.* **2022**, *63*, 245–251. [CrossRef] [PubMed]
18. Sluka, K.A.; Walsh, D. Transcutaneous electrical nerve stimulation: Basic science mechanisms and clinical effectiveness. *J. Pain* **2003**, *4*, 109–121. [CrossRef]
19. Ainsworth, L.; Budelier, K.; Clinesmith, M.; Fiedler, A.; Landstrom, R.; Leeper, B.J.; Moeller, L.; Mutch, S.; O'Dell, K.; Ross, J.; et al. Transcutaneous electrical nerve stimulation (TENS) reduces chronic hyperalgesia induced by muscle inflammation. *Pain* **2006**, *120*, 182–187. [CrossRef] [PubMed]
20. Desantana, J.M.; Sluka, K.A.; Lauretti, G.R. High and low frequency TENS reduce postoperative pain intensity after laparoscopic tubal ligation: A randomized controlled trial. *Clin. J. Pain* **2009**, *25*, 12–19. [CrossRef]
21. Liebano, R.E.; Vance, C.G.; Rakel, B.A.; Lee, J.E.; Cooper, N.A.; Marchand, S.; Walsh, D.M.; Sluka, K.A. Transcutaneous electrical nerve stimulation and conditioned pain modulation influence the perception of pain in humans. *Eur. J. Pain* **2013**, *17*, 1539–1546. [CrossRef]

22. Ferreira, A.P.; Costa, D.R.; Oliveira, A.I.; Carvalho, E.A.; Conti, P.C.; Costa, Y.M.; Bonjardim, L.R. Short-term transcutaneous electrical nerve stimulation reduces pain and improves the masticatory muscle activity in temporomandibular disorder patients: A randomized controlled trial. *J. Appl. Oral Sci.* **2017**, *25*, 112–120. [CrossRef] [PubMed]
23. Monaco, A.; Sgolastra, F.; Pietropaoli, D.; Giannoni, M.; Cattaneo, R. Comparison between sensory and motor transcutaneous electrical nervous stimulation on electromyographic and kinesio-graphic activity of patients with temporomandibular disorder: A controlled clinical trial. *BMC Musculoskelet. Disord.* **2013**, *14*, 168. [CrossRef] [PubMed]
24. Ortu, E.; Pietropaoli, D.; Adib, F.; Masci, C.; Giannoni, M.; Monaco, A. Electromyographic evaluation in children orthodontically treated for skeletal Class II malocclusion: Comparison of two treatment techniques. *Cranio* **2019**, *37*, 129–135. [CrossRef]
25. Liu, C.; Kaneko, S.; Soma, K. Effects of a mandibular lateral shift on the condyle and mandibular bone in growing rats. *Angle Orthod.* **2007**, *77*, 787–793. [CrossRef] [PubMed]
26. Sato, C.; Muramoto, T.; Soma, K. Functional lateral deviation of the mandible and its positional recovery on the rat condylar cartilage during the growth period. *Angle Orthod.* **2006**, *76*, 591–597. [CrossRef]
27. Du, B.L.; Li, J.N.; Guo, H.M.; Li, S.; Liu, B. The Effect of Functional Mandibular Shift on the Muscle Spindle Systems in Head-Neck Muscles and the Related Neurotransmitter Histamine. *J. Craniofac. Surg.* **2017**, *28*, 1628–1634. [CrossRef] [PubMed]
28. Goto, T.K.; Nishida, S.; Yahagi, M.; Langenbach, G.E.; Nakamura, Y.; Tokumori, K.; Sakai, S.; Yabuuchi, H.; Yoshiura, K. Size and orientation of masticatory muscles in patients with mandibular laterognathism. *J. Dent. Res.* **2006**, *85*, 552–556. [CrossRef]
29. Kasat, V.; Gupta, A.; Ladda, R.; Kathariya, M.; Saluja, H.; Farooqui, A.A. Transcutaneous electric nerve stimulation (TENS) in dentistry—A review. *J. Clin. Exp. Dent.* **2014**, *6*, e562–e568. [CrossRef]
30. Monaco, A.; Cattaneo, R.; Ortu, E.; Constantinescu, M.V.; Pietropaoli, D. Sensory trigeminal ULF-TENS stimulation reduces HRV response to experimentally induced arithmetic stress: A randomized clinical trial. *Physiol. Behav.* **2017**, *173*, 209–215. [CrossRef]
31. Moon, H.J.; Lee, Y.K. The relationship between dental occlusion/temporomandibular joint status and general body health: Part 1. Dental occlusion and TMJ status exert an influence on general body health. *J. Altern. Complement. Med.* **2011**, *17*, 995–1000. [CrossRef]
32. Zhulev, E.N.; Ershov, P.E.; Ershova, O.A. Influence of distal misalignment of the mandible on the development of postural pathology. *Stomatologiia* **2018**, *97*, 71–74. [CrossRef] [PubMed]
33. Ciarrocchi, I.; Ortu, E.; Masci, C.; Necozione, S.; Pietroapoli, D.; Monaco, A. Malocclusion and perinatal factors: A retrospective study. *Int. J. Clin. Exp. Med.* **2016**, *9*, 22758–22763.
34. Silva, A.P.; Sassi, F.C.; Andrade, C.R. Oral-motor and electromyographic characterization of patients submitted to open a nd closed reductions of mandibular condyle fracture. *Codas* **2016**, *28*, 558–566. [CrossRef]
35. Iodice, G.; Danzi, G.; Cimino, R.; Paduano, S.; Michelotti, A. Association between posterior crossbite, skeletal, and muscle asymmetry: A systematic review. *Eur. J. Orthod.* **2016**, *38*, 638–651. [CrossRef] [PubMed]
36. Tsanidis, N.; Antonarakis, G.S.; Kiliaridis, S. Functional changes after early treatment of unilateral posterior cross-bite associated with mandibular shift: A systematic review. *J. Oral Rehabil.* **2016**, *43*, 59–68. [CrossRef] [PubMed]
37. Baldini, A.; Nota, A.; Cravino, G.; Cioffi, C.; Rinaldi, A.; Cozza, P. Influence of vision and dental occlusion on body posture in pilots. *Aviat. Space Environ. Med.* **2013**, *84*, 823–827. [CrossRef] [PubMed]
38. Marchili, N.; Ortu, E.; Pietropaoli, D.; Cattaneo, R.; Monaco, A. Dental Occlusion and Ophthalmology: A Literature Review. *Open Dent. J.* **2016**, *10*, 460–468. [CrossRef]
39. Khan, M.T.; Verma, S.K.; Maheshwari, S.; Zahid, S.N.; Chaudhary, P.K. Neuromuscular dentistry: Occlusal diseases and posture. *J. Oral Biol. Craniofac. Res.* **2013**, *3*, 146–150. [CrossRef]
40. Miralles, R.; Gutierrez, C.; Zucchino, G.; Cavada, G.; Carvajal, R.; Valenzuela, S.; Palazzi, C. Body position and jaw posture effects on supra- and infrahyoid electromyographic activity in humans. *Cranio* **2006**, *24*, 98–103. [CrossRef]
41. Savastano, F. Applying neuromuscular techniques in the orthodontic setting. *S. Eur. J. Orthod. Dentofac.* **2017**, *4*, 31–42. [CrossRef]
42. Savastano, F. Contemporary Dental Occlusion in Orthodontics. *BAOJ Dent.* **2015**, *1*, 1.
43. Chan, C.A. A gneuromuscolar approach (part 1). *Dent. Asia Novemb. Dec.* **2015**, 39–42.
44. Awan, K.H.; Patil, S. The Role of Transcutaneous Electrical Nerve Stimulation in the Management of Temporomandibular Joint Disorder. *J. Contemp. Dent. Pract.* **2015**, *16*, 984–986. [CrossRef] [PubMed]
45. Monaco, A.; Cattaneo, R.; Marci, M.C.; Pietropaoli, D.; Ortu, E. Central Sensitization-Based Classification for Temporomandibular Disorders: A Pathogenetic Hypothesis. *Pain Res. Manag.* **2017**, *2017*, 5957076. [CrossRef]

applied sciences

MDPI

Article

Examining the Influence of Self-Esteem and Digital Literacy on Professional Competence Factors in Dental Education: A Cross-Sectional Study

Gulsum Ceylan [1,*], Melike Ozlem Eken [2], Selen Yuruk [2] and Faruk Emir [3]

[1] Department of Prosthodontics, School of Dentistry, Istanbul Medipol University, Istanbul 34083, Turkey
[2] Department of Prosthodontics, Institute of Health Sciences, Istanbul Medipol University, Istanbul 34810, Turkey
[3] Department of Prosthodontics, Gülhane Faculty of Dentistry, Health Sciences University, Ankara 06010, Turkey
* Correspondence: gulsumcyln@gmail.com; Tel.: +90-5336695745

Abstract: This study aims to investigate the effects of digital literacy and self-esteem on dental students' perceptions of professional competence and to provide recommendations for improving educational programs in dental schools. A mixed-methods, descriptive, and relational scanning design was employed, using a questionnaire comprising demographic information, the Rosenberg Self-Esteem Scale (RSES), the Digital Literacy Scale (DLS), and the Professional Competence Scale (PCS). The sample included 427 dental students in Istanbul, Turkey. Statistical analyses included Cronbach Alpha, Confirmatory Factor Analysis, Mann–Whitney U, Kruskal–Wallis, Spearman's rho correlation, and the Generalized Linear Model (Logit Model). The majority of participants were female (65.6%), aged 21 or older (85.2%), and in their 4th grade of study (35.2%). There were no significant differences in PCS scores concerning gender, age, grade, type of residence, and residence location ($p > 0.05$). However, PCS scores were significantly higher among those who listed dentistry among their top five preferences, expressed high satisfaction with education, and demonstrated high professional interest ($p < 0.05$). A significant, positive correlation was observed between PCS and RSES ($r = 0.398$; $p < 0.01$), DLS ($r = 0.404$; $p < 0.01$), preference ($r = 0.120$; $p < 0.05$), education satisfaction ($r = 0.298$; $p < 0.01$), and occupational interest ($r = 0.502$; $p < 0.05$). Furthermore, the Logit Model analysis revealed that RSES ($B = 0.290$; $p < 0.01$), DLS ($B = 0.258$; $p < 0.01$), education satisfaction ($B = 0.806$; $p < 0.05$), and occupational interest ($B = 3.825$; $p < 0.01$) significantly influenced PCS. The findings underscore self-esteem and digital literacy's integral role in shaping dental students' perceptions of their professional abilities. Digital literacy is a substantial pillar supporting students' professional competency. Coupled with a genuine interest in the field and bolstered self-confidence, digital literacy proves instrumental in enhancing students' academic success in dentistry. By emphasizing and expanding upon digital literacy content within the curriculum, universities can further propel advancements in the dental domain. As such, it is pivotal for dental institutions to weave strategies into their educational fabric to elevate students' self-esteem and digital proficiency.

Keywords: dental education; dentistry students; digital literacy; self-esteem

Citation: Ceylan, G.; Eken, M.O.; Yuruk, S.; Emir, F. Examining the Influence of Self-Esteem and Digital Literacy on Professional Competence Factors in Dental Education: A Cross-Sectional Study. *Appl. Sci.* **2023**, *13*, 9411. https://doi.org/10.3390/app13169411

Academic Editors: Giuseppe Minervini and Rocco Franco

Received: 16 June 2023
Revised: 15 August 2023
Accepted: 18 August 2023
Published: 19 August 2023

1. Introduction

Training in dentistry is critical for enhancing the quality of public health services within a country and promoting competitiveness in global health tourism [1,2]. Evaluating professional competence and determining influential factors during education is vital for future career success, as they significantly shape an individual's work life and contribute to overall well-being [3,4].

Self-esteem has been defined in various ways, making it somewhat challenging to define precisely. Essentially, it reflects how an individual feels about themselves. Two primary types of self-esteem are recognized: trait and state self-esteem. Trait self-esteem pertains to a person's consistent level of self-worth over time and across different situations. In contrast, state self-esteem deals with day-to-day variations in one's feelings of self-worth [5]. Self-esteem influences not only one's professional life but also their overall well-being [6,7]. Especially in professions like dentistry, where expertise and self-assurance are crucial [8,9], self-esteem can significantly impact professional competence.

In 2001, Prensky introduced the term "digital natives" to describe the generation born in and after 1980, who have grown up surrounded by digital devices and exhibit distinct learning styles from previous generations. Digital natives thrive in a culture of online connection, creativity, and sharing, living a digital existence that revolves around the internet [10]. They acquire information, communicate, engage in activities such as blogging, gaming, and social media interactions, and even perform transactions online. They favor visual information over text, prefer active learning, and can multitask and process information simultaneously [11].

With technology integrating into everyday life, digital literacy has become increasingly vital, closely linked with professional qualifications [12]. The recognition of digital literacy by various international organizations has led governments to revise their educational policies, legislation, and learning environments [13]. To implement 21st-century competencies, primarily digital competence, in educational public policy, a well-articulated plan is required [14].

Digital literacy is broadly characterized in the literature as the competencies and skills required to navigate our intricate and multifaceted information landscape, particularly when incorporating technology into academic courses [15]. At present, digital literacy encompasses three main pillars: (a) identifying and consuming digital content; (b) generating digital content; and (c) disseminating or distributing digital content. Various definitions of digital literacy exist, each intertwining different technical and non-technical facets [16].

Central to digital literacy is comprehending and decoding digital content [17,18]. This skill has become essential, especially since clinical research is predominantly available online, and digital tools are becoming standard for emerging research projects [19,20]. In the academic setting, students use digital technologies—ranging from diverse software to information hardware [21]—for numerous educational activities. These include accessing digital learning platforms, communicating via email, engaging with e-journals or e-books, and taking part in online evaluations [11].

In dentistry, precision, expertise, and patient interaction are fundamental. Self-esteem directly impacts professional performance, with confident professionals more likely to make informed decisions, communicate effectively with patients, and adapt to challenging clinical scenarios. As healthcare undergoes a digital revolution, digital literacy has become paramount, particularly as the reliance on digital tools for patient records, imaging, and procedures increases [22]. Understanding how self-esteem and digital literacy influence the perceived professional competence of dental students can provide valuable insights, leading to targeted strategies in dental education to enhance these aspects and ultimately improve patient care [23].

Given the importance of these elements in dental practice and the current gaps in understanding their role in shaping professional competence among dental students, it is imperative to further explore these interactions. This study aims to investigate the influence of dental students' digital literacy and self-esteem on their professional competence.

The potential findings from this study could significantly impact dental education and practice. By understanding the role and interplay of self-esteem and digital literacy in shaping professional competence, we can inform the design of educational strategies to foster these crucial attributes. This could result in a more competent and confident generation of dental professionals, enhance patient care, and contribute to improved oral health outcomes in the community.

2. Materials and Methods

2.1. Study Design and Sampling

This research utilized a mixed-methods design, incorporating descriptive and relational scanning models. The study aimed to shed light on specific phenomena or concepts through this integrated approach by leveraging qualitative and quantitative data sources. During the first semester of the 2022–2023 academic year, an online questionnaire was administered using Google Forms (Mountain View, CA, USA). This questionnaire comprised eight multiple-choice demographic questions, ten questions from the Rosenberg Self-Esteem Scale with scores ranging from 0 to 4, ten questions from the Digital Literacy Scale with scores ranging from 1 to 5, and eleven questions from the Professional Competence Scale, also scored between 1 and 5. The entire set, amounting to 39 questions, typically required around 15 min for completion. The survey exclusively contained closed-ended questions. Participants' ages varied from 18 to 31 years old. The utilized scales were sourced from established instruments with proven validity and reliability in Turkish.

Cohen et al. suggest that for a descriptive survey model, having at least 384 participants yields a 95% confidence interval and a 0.05 significance level when aiming to represent a population ranging from one to five million individuals [24].

Building on findings by Yu et al. [9], the effect size was determined to be 0.3097572 using G-Power 3.1.9.2. Given this effect size, a 95% confidence interval, and a 0.05 deviation level, the study required a minimum of 115 participants. However, this research gathered data from 427 dentistry students in Istanbul, offering a comprehensive perspective from both private and public institutions. This cross-sectional approach included students from all academic levels within the dentistry program. Participants were selected through a straightforward random sampling method based on their willingness to participate.

2.2. Measuring Tools

The questionnaire form includes a demographic information form, Self-Esteem Scale, Digital Literacy Scale, and Vocational Competence Scale.

2.2.1. Demographic Information Form

This form has been meticulously designed to gather a comprehensive understanding of the student participants' backgrounds. It solicits information about various facets of their personal and academic lives. Specifically, the form inquires about the student's gender and age, which offers insights into the diversity of the sample. Further, it delves into academic specifics, such as the student's current grade level and the reason behind their department choice. The form also seeks to understand their living situation, capturing details on their residence type and living arrangements before joining their current department. Additionally, it gauges their overall satisfaction and engagement level in their chosen field by inquiring about their level of education within the department and their genuine interest in pursuing the profession. Such detailed demographic data provides a richer context to interpret the results of the study.

2.2.2. Rosenberg Self-Esteem Scale (RSES)

The RSES, devised by Rosenberg and later adapted for Turkish audiences by Çuhadaroğlu, is a prominent tool for gauging self-esteem [25]. With its 10 items evaluated on a 4-point Likert scale, the RSES sheds light on an individual's sense of self-worth and feelings about themselves. Its reputation for reliability is confirmed by a previously reported Cronbach Alpha value of 0.81. Within the framework of this research, the scale showcased its consistent reliability with a Cronbach's Alpha of 0.882. Its extensive application and proven reliability highlight its significance as a trusted measure in psychological and sociological studies.

2.2.3. Digital Literacy Scale (DLS)

The DLS, initially crafted by Ng and subsequently adapted for Turkish audiences by Ustundag et al., offers a comprehensive assessment of digital literacy [10,26]. The tool consists of 10 items evaluated on a 5-point Likert scale, encapsulating the unified dimension of digital proficiency. Historically, the DLS has demonstrated strong internal consistency, as evidenced by a Cronbach Alpha value of 0.86. In the context of this particular study, the instrument exhibited even higher reliability, achieving a Cronbach's Alpha of 0.894. This underscores the scale's robustness and ability to consistently gauge respondents' digital capabilities in varied settings.

2.2.4. Professional Competence Scale (PCS)

Based on expert opinions and the relevant literature, the Professional Competence Scale aims to measure perceptions of vocational aptitude. The scale comprises 11 items, utilizing a 5-point Likert-type format to gauge responses, ranging from "Strongly Disagree" to "Strongly Agree" [27,28]. Participants' total scores can range from 11 to 55, with a higher score indicating a more favorable perception of one's professional competence. The scale has demonstrated both reliability and validity in various settings. In this study, the PCS exhibited internal solid consistency with a Cronbach's Alpha value of 0.851, showcasing its appropriateness for assessing dental students' perceptions of their professional abilities.

2.3. Ethical Considerations

This study strictly adhered to ethical standards and received approval from the Ethics Committee. Before launching the cross-sectional study, participants were presented with a consent form detailing the research objectives. Their participation was entirely based on their discretion, ensuring it was voluntary. The study refrained from collecting personal identifiers to maintain confidentiality and privacy.

2.4. Statistical Analysis

Demographic details were presented using frequencies. For scale scores, means and standard deviations were the primary descriptors. Reliability was ensured through Cronbach's Alpha for internal consistency, while the Confirmatory Factor Analysis was employed to validate the data. The Kolmogorov–Smirnov test assessed the normality distribution of the study's parameters. Given that the distribution of all scale means was non-normal, nonparametric tests became the approach. The Mann–Whitney U test was chosen to highlight differences between the two groups. For identifying distinctions across more than two groups, the Kruskal–Wallis test was implemented.

Relationship analyses utilized Spearman's rho correlation for univariate nonparametric correlations and the Generalized Linear Model (Logit Model) for multivariate nonparametric regression. The Generalized Linear Model is specially designed for nonparametric regressions, making it different from typical logit models. One advantage of this model is its ability to work directly with scale parameters without strictly dividing the data into two or multiple distinct categories. However, all regression models showed some deviations [29]. With its logit assumption, the Generalized Linear Model effectively reduced deviations for nonparametric variables. All statistical analyses were conducted using the SPSS 25.0 software, with a confidence level set at 95% and a significance threshold of 0.05.

3. Results

The research encapsulates a detailed exploration of dental students' demographic and academic inclinations. Predominantly, females formed a significant segment, representing 65.6% of the study's participants. Dovetailing with this, a remarkable 85.2% were aged 21 or older, harmonizing with the observation that 35.2% were engrossed in their fourth year of academic pursuit. Delving into the socio-cultural milieu, 40% of these budding professionals cohabited with family members. Moreover, Istanbul, the city under study, was a new home for 57.8% of these students, having migrated from other provinces. In the academic realm, 57.6% expressed a definitive alignment with their current department, marking it their top academic choice.

Transitioning to their academic perceptions, the landscape is slightly variegated. While a measurable 55.7% assessed their departmental educational quality as moderate, a near equivalent 48.5% articulated an elevated zeal for their academic discipline (Table 1).

Table 1. Baseline characteristics of participants.

		Count	Percentage (%)
Gender	Female	280	65.6
	Male	147	34.4
Age	20 and above	63	14.8
	21 and higher	364	85.2
Grade	1st grade	2	0.5
	2nd grade	82	19.2
	3rd grade	81	19.0
	4th grade	150	35.2
	5th grade	111	26.1
Residence type	Dormitory	69	16.2
	With friends	120	28.1
	With family	171	40.0
	Other	67	15.7
Preference	First five	246	57.6
	Second five	108	25.3
	Other	73	17.1
Residence location	Istanbul	180	42.2
	Outside	247	57.8
Education evaluation	Very little	5	1.2
	Little	23	5.4
	Moderate	238	55.7
	High	140	32.8
	Very high	21	4.9
Occupation interest	Very little	3	0.7
	Little	12	2.8
	Moderate	97	22.7
	High	207	48.5
	Very high	108	25.3

Upon assessing the Professional Competence Score (PCS), an intricate exploration revealed that several factors, such as gender, age, academic tenure, residential category, and geographic origin, did not yield any discernible statistical divergence ($p > 0.05$). Contrastingly, students exuding high academic satisfaction and those ranking their department in the top echelons showcased significantly elevated PCS scores ($p < 0.05$) (Table 2).

Table 2. PCS score differences according to the demographic properties of participants.

		Mean	Std. Deviation	Test Value	*p* Value
Gender	Female	42.00	7.20	18,449.500 (U)	0.078 [a]
	Male	40.67	7.24		
Age	20 and above	42.30	6.66	10,832.500 (U)	0.483 [a]
	21 and higher	41.41	7.33		
Grade	1st grade	31.50	2.12	8.297 (X^2)	0.081 [b]
	2nd grade	42.87	6.85		
	3rd grade	42.11	7.05		
	4th grade	40.86	7.43		
	5th grade	41.22	7.27		

Table 2. *Cont.*

		Mean	Std. Deviation	Test Value	*p* Value
Residence type	Dormitory	42.43	7.10	6.296 (X^2)	0.098 [b]
	With friends	40.09	7.56		
	With family	42.17	7.41		
	Other	41.64	5.98		
Preference	First five	42.26	7.29	6.196 (X^2)	0.045 [b]
	Second five	40.83	7.19		
	Other	40.18	6.89		
Residence location	Istanbul	42.17	7.14	20,687.500 (U)	0.220 [a]
	Outside	41.09	7.28		
Education satisfaction	Very little	34.80	9.55	39.819 (X^2)	0.000 [b]
	Little	36.61	7.74		
	Moderate	40.56	7.07		
	High	43.44	6.58		
	Very high	47.10	5.49		
Occupation interest	Very little	28.67	11.24	110.891 (X^2)	0.000 [b]
	Little	30.33	6.33		
	Moderate	36.80	6.98		
	High	42.29	5.94		
	Very high	45.97	5.50		

[a]. Mann–Whitney U Test (U); [b]. Kruskal–Wallis Test (X^2).

Spearman's rho correlation analysis further accentuates the statistical robustness of the study. This methodological tool substantiated that PCS shared a positive synergy with self-esteem (RSES, r = 0.398; $p < 0.01$), digital literacy (DLS, r = 0.404; $p < 0.01$), departmental preference (r = 0.120; $p < 0.05$), satisfaction with the quality of education (r = 0.298; $p < 0.01$), and an intrinsic occupational fervor (r = 0.502; $p < 0.05$) (Table 3).

Table 3. Spearman's rho correlation analysis results between PCS, SES, DLS scores, and significantly different parameters for PCS scores.

	r	*p*
RSES Total	0.398 **	0.000
DLS Total	0.404 **	0.000
Preference	0.120 *	0.013
Education satisfaction	0.298 **	0.000
Occupation interest	0.502 **	0.000

* $p < 0.05$ ** $p < 0.01$ RSES: Rosenberg Self-Esteem Scale; DLS: Digital Literacy Scale; PCS: Professional Competence Scale.

A subsequent evaluation using the Generalized Linear Model (Logit Model) further buttressed these findings, unequivocally emphasizing the statistical significance of variables such as RSES (B = 0.290; $p < 0.01$), DLS (B = 0.258; $p < 0.01$), educational contentment (B = 0.806; $p < 0.05$), and notably, professional zeal (B = 3.825; $p < 0.01$) in influencing PCS. (Table 4).

Table 4. Generalized linear model (logit model) analysis results for correlated factors with PCS.

Parameter	B	Std. Error	95% Wald Confidence Interval		Hypothesis Test	
			Lower	Upper	Wald X^2	p
(Intercept)	9.298	1.82	5.74	12.86	26.221	0.000
[Preference = First five]	1.199	0.70	−0.18	2.58	2.913	0.088
[Preference = Second five]	0.000	0.79	−1.56	1.56	0.000	>0.05
[Preference = Other]	0 [a]	.	.	.	.	.
RSES Total	0.290	0.05	0.20	0.38	38.193	0.000
DLS Total	0.258	0.03	0.20	0.33	53.396	0.000
Education satisfaction	0.806	0.38	0.05	1.56	4.399	0.036
Occupation interest	3.825	0.33	3.17	4.48	130.654	0.000
(Scale)	27.478 [b]	1.88	24.03	31.42		

Dependent Variable: PCS Total; Model: (Intercept), Preference, SES_Total, DLS_Total, Education satisfaction, Occupation interest; [a]. Set to zero because this parameter is redundant; [b]. Maximum likelihood estimate; RSES: Rosenberg Self-Esteem Scale; DLS: Digital Literacy Scale; PCS: Professional Competence Scale.

The core findings of this study highlight the critical role of both self-esteem and digital literacy in shaping dental students' perceptions of professional competence. Notably, significant positive correlations were observed between self-esteem and professional competence (RSES: r = 0.398; $p < 0.01$), as well as between digital literacy and professional competence (DLS: r = 0.404; $p < 0.01$). In addition, professional interest (r = 0.502; $p < 0.05$) and education satisfaction (r = 0.298; $p < 0.01$) showed significant positive correlations with professional competence. The analysis also revealed that the influence of professional interest on professional competence (B = 3.825; $p < 0.01$) was the most robust among the factors evaluated. It is important to note that these relationships held even after adjusting for other influential factors in the Logit model analysis.

These findings show that self-esteem and digital skills play a significant role in shaping how dental students think about their professional abilities. Impressively, the students' genuine enthusiasm for their field is the main factor, even when set against other elements in the Logit model analysis. The data therein underscore the indispensability of both self-esteem and digital literacy in shaping and potentially enhancing professional competence perceptions among dental students.

4. Discussion

This study investigated the impact of digital literacy and self-esteem on dental students' perceptions of their professional competence. A questionnaire was administered to 427 dentistry students, and the collected data were analyzed. The results demonstrated a positive correlation between digital literacy and self-esteem with students' perceived professional competence.

The literature on professional competence presents varying findings. Some studies report that professional competence differs according to demographic and social characteristics [30–32], while others propose the contrary [33]. In single-centered studies, where demographic characteristics are generally similar, the perception of professional competence does not vary based on these factors. However, in multicenter studies, professional competence levels significantly differ based on demographic characteristics [34–36].

In this study, the perception of professional competence significantly differed only in relation to the order of preference. It is noteworthy that this research was conducted on a specific group of dentistry students at a single center, which may have influenced the results. Different outcomes might be found in various departments or occupational groups and larger samples or multicenter studies.

Factors affecting the perception of professional competence include an individual's professional characteristics and interest in their chosen profession. Individuals who choose their profession willingly and have a high level of interest tend to possess higher professional competence and experience greater life satisfaction [37–39]. In line with this, both educational satisfaction and interest in the profession were found to positively and significantly impact professional competence. Atalayin et al. found that dentistry students residing away from their hometowns exhibited lower academic proficiency compared to those living in their native cities [40]. However, in this study, the educational success of students from other cities was statistically similar to those who resided in Istanbul.

Understanding the factors influencing one's job decision is crucial because it shapes future expectations and could influence career satisfaction. While some people join the workforce with substantial knowledge and reasonable expectations, others might end up in jobs they have yet to learn. Some people choose their careers based on the strategy of least resistance, such as pursuing a career path that their parents have recommended or imitating an older sibling [40]. According to the results of our study, people who choose a job consciously and voluntarily tend to have higher educational performance and self-esteem. Research findings reveal a pronounced positive correlation between self-esteem levels in individuals who placed dentistry among their top five university preferences and those with very high occupational interests. Consistent with these results, a systematic review also identified a significant positive relationship between professional awareness and self-esteem [5].

While the literature contains studies on digital literacy and self-esteem, research examining the effects of these two factors on professional competence is scarce [41,42]. This study found that both digital literacy and self-esteem positively influenced professional competence, with self-esteem having a greater impact.

The findings of this study highlight a positive correlation between students' self-esteem and their educational success. Enhanced psychological well-being can elevate self-esteem among students. This observation aligns with prior research underscoring the pivotal role of psychological health in influencing self-esteem [43,44]. Moreover, it was found that students' positive emotions are directly linked to their self-esteem. Students experiencing positive feelings often maintain a robust and favorable self-esteem, affirming their self-worth and recognizing the respect they merit [45].

Digital literacy also augments one's esthetic perception [46]. Consequently, digital literacy might bolster the esthetic judgments of dentists and dental students engaged in aesthetic dental procedures, amplifying their professional self-assuredness.

Dozic et al. [47] noted a preference among dental students for digital systems over visual and conventional ones, suggesting a shift toward digital literacy. In a separate study, Shooriabi and Gilavand [48] observed that dental students in Iran frequently employ smartphones and digital media for educational endeavors. Similarly, Rung et al. [49] reported dental students' prevalent use of digital and social media for educational activities in Australia.

The incorporation of smartphones as a staple in daily life and as an adjunct tool in many professions is especially relevant in the realm of dentistry. Dentists, predominantly from the digital native generation, can capitalize on the multifaceted utilities presented by smartphones, such as capturing visuals and facilitating text exchanges with patients during diagnostic and treatment phases [50].

A key limitation of this study, and others in the field, is the lack of a specific scale for dentistry, despite the proven validity and reliability of the scales used. Additionally, the single-centered nature of the sample with a similar demographic structure is another notable limitation.

The main contribution of this research to the literature and the field lies in its applicability and pragmatism. By incorporating self-perception and digital literacy education into university curricula, high professional gains can be achieved at very low costs. Another contribution is the multi-disciplinary nature of the study, which covers both professional development and dentistry.

Given the significant impact of self-esteem and digital literacy on dental students' professional competence, further research in these areas is warranted. Future studies might explore specific interventions to enhance these factors among dental students and their impact on professional competence. Moreover, expanding the research to include students from various demographic backgrounds and multiple centers would provide a more comprehensive understanding.

The implications of this study extend beyond dental education, potentially influencing the broader field of healthcare education and practice. Given the significant correlations between self-esteem, digital literacy, and perceptions of professional competence, these findings underscore the need to prioritize these factors within the curriculum across healthcare disciplines. By incorporating strategies to enhance self-esteem and digital literacy, future healthcare professionals can be better prepared to adapt to the increasingly digital healthcare environment. Thus, these findings underscore the importance of focusing on holistic development in healthcare education.

5. Conclusions

The findings of this study highlight the significant influence of both self-esteem and digital literacy on dental students' perceptions of professional competence. Recognizing these factors, dental schools should integrate methods into their curriculum that boost students' self-esteem and digital skills. By doing this, we can enhance professional competence and attract more students to the field, even those who might have initially shown less enthusiasm.

In today's fast-changing environment where healthcare meets digital technology, the importance of self-esteem and digital literacy cannot be understated. They play a crucial role in preparing the next generation of skilled dental practitioners. Therefore, it is vital for educational institutions to consider these insights when shaping the future of dental education and practice.

Overall, the research emphasizes that digital literacy greatly impacts students' professional competence. This impact, combined with a student's passion for the profession and their confidence, creates a better educational journey for dentistry students. Increasing focus on digital literacy within universities can positively change the domain.

Author Contributions: Conceptualization, M.O.E. and S.Y.; data curation, M.O.E.; formal analysis, G.C. and M.O.E.; investigation, M.O.E. and S.Y.; methodology, G.C. and M.O.E.; software, F.E.; supervision, G.C.; validation, F.E.; visualization, G.C. and M.O.E.; writing—original draft, G.C., M.O.E., and S.Y.; writing—review and editing, G.C. and F.E. All authors have read and agreed to the published version of the manuscript.

Funding: This research received no external funding.

Institutional Review Board Statement: The study was conducted according to the guidelines of the Declaration of Helsinki and approved by the Ethics Committee of Istanbul Medipol University (approval number: 2022/820).

Informed Consent Statement: Informed consent was obtained from all subjects involved in the study.

Data Availability Statement: The data presented in this study are available upon request from the corresponding author.

Acknowledgments: The authors thank all the participants who participated in this study.

Conflicts of Interest: The authors declare no conflict of interest.

References

1. Cheng, F.C.; Wang, L.H.; Ozawa, N.; Chang, J.Y.F.; Liu, S.Y.; Chiang, C.P. Development of dental education for medical students in Taiwan during the Japanese colonial period. *J. Dent. Sci.* **2022**, *17*, 903–912. [CrossRef] [PubMed]
2. Farrokhi, F.; Mohebbi, S.Z.; Farrokhi, F.; Khami, M.R. Impact of COVID-19 on dental education-a scoping review. *BMC Med. Educ.* **2021**, *21*, 587. [CrossRef] [PubMed]
3. Widodo, W.; Gustari, I.; Chandrawaty, C. Adversity Quotient Promotes Teachers' Professional Competence More Strongly Than Emotional Intelligence: Evidence from Indonesia. *J. Intell.* **2022**, *10*, 44. [CrossRef] [PubMed]
4. Hanlon, H.R.; Prihodova, L.; Hoey, H.; Russell, T.; Donegan, D.; O'Shaughnessy, A. Attitudes, perceived benefits, and experiences of engagement with professional competence schemes for doctors in Ireland: Findings from a national survey. *J. Contin. Educ. Health Prof.* **2021**, *41*, 176–184. [CrossRef] [PubMed]
5. Randal, C.; Pratt, D.; Bucci, S. Mindfulness and self-esteem: A systematic review. *Mindfulness* **2015**, *6*, 1366–1378. [CrossRef]
6. Van der Aar, L.P.E.; Peters, S.; Becht, A.; Crone, E. Better self-concept, better future choices? Behavioral and neural changes after a naturalistic self-concept training program for adolescents. *Cogn. Affect. Behav. Neurosci.* **2022**, *22*, 341–361. [CrossRef] [PubMed]
7. Casino-García, A.M.; Llopis-Bueno, M.J.; Llinares-Insa, L.I. Emotional intelligence profiles and self-esteem/self-concept: An analysis of relationships in gifted students. *Int. J. Environ. Res. Public Health* **2021**, *18*, 1006. [CrossRef] [PubMed]
8. Wójcik, D.; Kutnik, J.; Szalewski, L.; Borowicz, J. Predictors of stress among dentists during the COVID-19 epidemic. *Sci. Rep.* **2022**, *12*, 7859. [CrossRef]
9. Yu, W.; Qian, Y.; Abbey, C.; Wang, H.; Rozelle, S.; Stoffel, L.A.; Dai, C. The Role of Self-Esteem in the Academic Performance of Rural Students in China. *Int. J. Environ. Res. Public Health* **2022**, *19*, 13317. [CrossRef]
10. Ng, W. Can we teach digital natives digital literacy? *Comput. Educ.* **2012**, *59*, 1065–1078. [CrossRef]
11. Jones, C.; Ramanau, R.; Cross, S.; Healing, G. Net generation or Digital Natives: Is there a distinct new generation entering university? *Comput. Educ.* **2010**, *54*, 722–732. [CrossRef]
12. Nguyen, L.A.T.; Habók, A. Tools for assessing teacher digital literacy: A review. *J. Comput. Educ.* **2023**, 1–42. [CrossRef]
13. Pérez-Escoda, A.; García-Ruiz, R.; Aguaded, I. Dimensions of digital literacy based on five models of development [Dimensiones de la alfabetización digital a partir de cinco modelos de desarrollo]. *Cult. Educ.* **2019**, *31*, 232–266. [CrossRef]
14. Chu, S.K.W.; Reynolds, R.B.; Tavares, N.J.; Notaria, M.; Lee, C.W.Y. Twenty-First Century Skills. In *Development through Inquiry-Based Learning from Theory to Practice*; Springer: Singapore, 2017; pp. 17–32.
15. Blau, I.; Shamir-Inbal, T.; Avdiel, O. How does the pedagogical design of a technology-enhanced collaborative academic course promote digital literacies, self-regulation, and perceived learning of students? *Internet High. Educ.* **2020**, *45*, 100722. [CrossRef]
16. Tinmaz, H.; Lee, Y.T.; Fanea-Ivanovici, M.; Baber, H. A systematic review on digital literacy. *Smart Learn. Environ.* **2022**, *9*, 21. [CrossRef]
17. Lindberg, V.; Jounger, S.L.; Christidis, M.; Christidis, N. Literacy as part of professional knowing in a Swedish dental education. *BMC Med. Educ.* **2021**, *21*, 373. [CrossRef] [PubMed]
18. Savić Pavičin, I.; Jonjić, A.; Maretić, I.; Dumančić, J.; Zymber Çeshko, A. Maintenance of dental records and forensic odontology awareness: A survey of Croatian dentists with implications for dental education. *Dent. J.* **2021**, *9*, 37. [CrossRef]
19. Kesici, A. The Effect of Digital Literacy on Creative Thinking Disposition: The Mediating Role of Lifelong Learning Disposition. *J. Learn. Teach. Digit. Age* **2022**, *7*, 260–273. [CrossRef]
20. Barrot, J.S.; Llenares, I.I.; Del Rosario, L.S. Students' online learning challenges during the pandemic and how they cope with them: The case of the Philippines. *Educ. Inf. Technol.* **2021**, *26*, 7321–7338. [CrossRef]
21. Mohammadyari, S.; Singh, H. Understanding the effect of e-learning on individual performance: The role of digital literacy. *Comput. Educ.* **2015**, *82*, 11–25. [CrossRef]
22. Orsini, C.A.; Binnie, V.I.; Tricio, J.A. Motivational profiles and their relationships with basic psychological needs, academic performance, study strategies, self-esteem, and vitality in dental students in Chile. *J. Educ. Eval. Health Prof.* **2018**, *15*, 11. [CrossRef] [PubMed]
23. Radeef, A.S.; Faisal, G.G. Low self-esteem and its relation with psychological distress among dental students. *Eur. J. Med. Health Sci.* **2019**, *1*, 1. [CrossRef]
24. Cohen, L.; Manion, L.; Morrison, K. The ethics of educational and social research. In *Research Methods in Education*, 8th ed.; Routledge: London, UK, 2017; pp. 111–143.
25. Cecen-Erogul, A.R. Psychometric properties of Turkish version of Childhood Trauma Questionnaire among adolescents with gender differences. *Psychology* **2012**, *3*, 916–922.
26. Ustundag, M.T.; Gunes, E.; Bahcivan, E. Turkish adaptation of digital literacy scale and investigating pre-service science teachers' digital literacy. *J. Educ. Future* **2017**, *12*, 19–29.
27. Zahoor, F.; Jumani, N.B.; Malik, S. Professional Qualifications and Competencies of Teacher Educators and Subject Teachers of Education: Gender Wise Analysis. *Glob. Bus. Rev.* **2019**, *4*, 158–167. [CrossRef]
28. Mak, B. Professional Qualifications of Teachers for English for Primary and Secondary Education—A Brief Comparison between Hong Kong and China. *J. Pan-Pac. Assoc. Appl. Linguist.* **2016**, *20*, 19–29.
29. Yılmaz, K.; Turanlı, M. A Multi-disciplinary Investigation of Linearization Deviations in Different Regression Models. *Asian J. Probab. Stat.* **2023**, *22*, 15–19. [CrossRef]

30. Turan, C.; Akın, Y. Investigation of the Relationship Between Socio-Demographical Characteristics and Organizational Factors Affecting Academic Achievement by Non-Linear Canonical Correlation Analysis: The Case of Trakya University Vocational School Students. *Soc. Sci. Res. J.* **2019**, *8*, 146–163.
31. Lasauskiene, J.; Rauduvaite, A. Expression of pre-service teachers' emotional competency in their educational practice. *Procedia Soc. Behav. Sci.* **2015**, *205*, 103–109. [CrossRef]
32. Kuisma, M.; Sandberg, A. Preschool teachers' and student preschool teachers' thoughts about professionalism in Sweden. *Eur. Early Child. Educ. Res. J.* **2008**, *16*, 186–195. [CrossRef]
33. Erişen, Y.; Celikoz, N. Efficacy Perceptions of Teacher Candidates on General Teaching Behaviors. *J. Turk. Sci. Educ.* **2003**, *1*, 427–439.
34. Song, M.; Choi, H.J.; Hyun, S.S. MBTI personality types of Korean cabin crew in Middle Eastern Airlines, and their associations with cross-cultural adjustment competency, occupational competency, coping competency, mental health, and turnover intention. *Int. J. Environ. Res. Public Health* **2021**, *18*, 3419. [CrossRef] [PubMed]
35. Tan, B.L.; Zhen Lim, M.W.; Xie, H.; Li, Z.; Lee, J. Defining occupational competence and occupational identity in the context of recovery in schizophrenia. *Am. J. Occup. Ther.* **2020**, *74*, 7404205120p1–7404205120p11. [CrossRef] [PubMed]
36. Lester, S.; Koniotaki, A.; Religa, J. ComProCom: A revised model of occupational competence. *Educ. Train.* **2018**, *60*, 290–302. [CrossRef]
37. Forsman, H.; Jansson, I.; Leksell, J.; Lepp, M.; Sundin Andersson, C.; Engström, M.; Nilsson, J. Clusters of competence: Relationship between self-reported professional competence and achievement on a national examination among graduating nursing students. *J. Adv. Nurs.* **2020**, *76*, 199–208. [CrossRef] [PubMed]
38. Li, L.; Li, G.; Chen, J. Professional competence or personal relationship? Research on the influencing mechanism on repeated purchase intention of agricultural resources. *Int. J. Environ. Res. Public Health* **2020**, *17*, 2278. [CrossRef]
39. Cheng, X.; Chen, J. An exploration of medical education in central and southern China: Measuring the professional competence of clinical undergraduates. *Int. J. Environ. Res. Public Health* **2019**, *16*, 4119. [CrossRef]
40. Atalayin, C.; Balkis, M.; Tezel, H.; Onal, B.; Kayrak, G. The prevalence and consequences of burnout on a group of preclinical dental students. *Eur. J. Dent.* **2015**, *9*, 356–363. [CrossRef]
41. Aguiar, C.M.; Pessoa, M.A.V.; Câmara, A.C.; Perrier, R.A.; de Figueiredo, J.A.P. Factors involved in the choice of dentistry as an occupation by Pernambuco dental students in Brazil. *J. Dent. Educ.* **2009**, *73*, 1401–1407. [CrossRef]
42. Kahveci, P. Language Teachers' Digital Literacy and Self-efficacy: Are They Related? *ELT Res. J.* **2021**, *10*, 123–139.
43. Dogan, T.; Totan, T.; Sapmaz, F. The role of self-esteem, psychological well-being, emotional self-efficacy, and affect balance on happiness: A path model. *Eur. Sci. J.* **2013**, *9*, 31–42.
44. Sarkova, M.; Bacikova-Sleskova, M.; Madarasova Geckova, A.; Katreniakova, Z.; Van den Heuvel, W.; Van Dijk, J.P. Adolescents' psychological well-being and self-esteem in the context of relationships at school. *Educ. Res.* **2014**, *56*, 367–378. [CrossRef]
45. Tran, M.A.Q.; Vo-Thanh, T.; Soliman, M.; Khoury, B.; Chau, N.N.T. Self-compassion, mindfulness, stress, and self-esteem among Vietnamese university students: Psychological well-being and positive emotion as mediators. *Mindfulness* **2022**, *13*, 2574–2586. [CrossRef] [PubMed]
46. Martínez-Bravo, M.C.; Sádaba Chalezquer, C.; Serrano-Puche, J. Dimensions of Digital Literacy in the 21st Century Competency Frameworks. *Sustainability* **2022**, *14*, 1867. [CrossRef]
47. Dozic, A.; Kharbanda, A.K.; Kamell, H.; Brand, H.S. European dental students' opinions about visual and digital tooth colour determination systems. *J. Dent.* **2011**, *39*, e23–e28. [CrossRef]
48. Shooriabi, M.; Gilavand, A. Investigating the use of smartphones for learning purposes by Iranian dental students. *World Fam. Med. J.* **2017**, *99*, 1–6. [CrossRef]
49. Rung, A.; Warnke, F.; Mattheos, N. Investigating the use of smartphones for learning purposes by Australian dental students. *JMIR mHealth uHealth* **2014**, *2*, e3120. [CrossRef]
50. Pourdanesh, F.; Sayyedi, A.; Jamilian, A.; Yaghmaei, M. Application of self-recorded photos using mobile phones in maxillofacial surgery. *J. Mob. Technol. Med.* **2012**, *1*, 46–49. [CrossRef]

applied sciences

MDPI

Article

Quantitative Evaluation of the Infrazygomatic Crest Thickness in Polish Subjects: A Cone-Beam Computed Tomography Study

Marta Gibas-Stanek *, Julia Ślusarska, Michał Urzędowski, Szczepan Żabicki and Małgorzata Pihut

Department of Prosthodontics and Orthodontics, Dental Institute, Faculty of Medicine, Jagiellonian University Medical College, Montelupich St. 4/108, 31-155 Krakow, Poland; jula.slusarska@gmail.com (J.Ś.); michal.urzedowski@gmail.com (M.U.); szczepanzabicki@gmail.com (S.Ż.); malgorzata.pihut@uj.edu.pl (M.P.)
* Correspondence: marta.gibas-stanek@uj.edu.pl

Abstract: Infrazygomatic crest (IZC) mini-implants are frequently used as an absolute anchorage when intrusive or distally directed forces are required during orthodontic treatment. The aim of the present study was to evaluate the thickness of the IZC area in Polish patients as well as to assess dependency between bone availability, sex, and age. The study material was 100 cone beam computed tomography scans (CBCT) of the maxilla of patients of the University Dental Clinic in Krakow (50 men and women each). IZC bone thickness was measured at nine different points. The biggest bone thickness was recorded in the interdental space between the first and second molar at the height of 12 mm (6.03 ± 2.64 mm). The thinnest bone depth was localized at the level of the mesial root of the first molar, 16 mm above the occlusal plane (2.42 ± 2.16). There was a significant and negative correlation between bone thickness and age in the case of measurements taken buccally to the first molar. Only two out of nine measurements showed a sex dependency (points I2 and I3). Considering vertical and sagittal dimensions, the most favorable conditions for IZC mini-implant placement were found interdentally, between the first and second molar, 12 mm above the occlusal plane.

Keywords: bone screw; maxilla; orthodontics; X-ray computed tomography

Citation: Gibas-Stanek, M.; Ślusarska, J.; Urzędowski, M.; Żabicki, S.; Pihut, M. Quantitative Evaluation of the Infrazygomatic Crest Thickness in Polish Subjects: A Cone-Beam Computed Tomography Study. *Appl. Sci.* **2023**, *13*, 8744. https://doi.org/10.3390/app13158744

Academic Editors: Giuseppe Minervini and Rocco Franco

Received: 25 June 2023
Revised: 23 July 2023
Accepted: 25 July 2023
Published: 28 July 2023

1. Introduction

The infrazygomatic crest (IZC) is a thick bony ridge located between the first and the second maxillary molar. It has been used as a site for orthodontic miniplate placement when absolute anchorage or intrusive or distally directed forces were required during orthodontic treatment [1,2]. Hugo de Clerck, the author of the method, recommends this location due to its solid bone structure and safe distance from the roots of the upper molars [1]. The Zygoma Anchor System, designed and popularized by de Clerck, consists of a titanium miniplate adjusted to the shape of the infrazygomatic buttress and three miniscrews (5 or 7 mm long), that fixes the miniplate to the bone. Although miniplates provide effective orthodontic anchorage, the process of their placement requires advanced surgical skills. The procedure consists of preparation of the mucoperiosteal flap, bending of the miniplate to the proper shape, fixation with miniscrews, repositioning of the mucoperiosteum, and placing sutures. After completion of the orthodontic treatment, miniplates have to be removed. To simplify the clinical procedure and minimize the financial cost of orthodontic treatment, extra-alveolar IZC miniscrews have been introduced. They are placed on the buccal surface of the alveolar process at the base of the zygomatic crest eminence as an alternative to miniplates. Chris Chang, who popularized IZC, recommends its use for the retraction of posterior teeth and rotation of the whole dental arch in order to manage even challenging malocclusions without extractions or orthognathic surgery [3].

Despite their undeniable advantages, extra-alveolar miniscrews inserted in the IZC area can perforate the maxillary sinus and initiate sinus infection or mini-implant loss [4]. Although according to some authors [5], sinus penetration depth within 1 mm is advocated for mini-implant anchorage, to avoid this complication, adequate bone depth should be available at the site of mini-screw insertion. Typically, IZC mini-screws are longer than intra-radicular mini-implants and manufactured in two dimensions: 2 mm diameter and 12 mm or 14 mm length (a 14 mm screw is recommended in the case of thick soft tissue in the buccal vestibule) [6]. As stated by the literature, a minimum of 6 mm of the bone is necessary to provide sufficient stability of the IZC during orthodontic treatment [7,8]. According to our clinical observations of Polish patients, in the vast majority of cases, the depth of bone ridge in this area is thinner, making IZC mini-screw placement challenging or even impossible. In the face of a lack of studies assessing the maxillary bone thickness for optimum IZC mini-screw placement in the Polish population, the aim of the present study was to evaluate the thickness of the IZC area.

It was found that the IZC is located between the maxillary second premolar and the first molar in youngs and at the level of the maxillary first molar in adults [9]. Liou developed a method for IZC screw placement adjacent to buccal surfaces of the maxillary first molar [7], while Lin recommends a more distal site that is buccal to the maxillary second molars [10,11]. Nevertheless, the region between the first and the second molar is also recommended for IZC screw placement [12]. To localize the most favorable site for IZC mini-screw implantation, we compared bone thickness at different levels considering sagittal plane and vertical distance from the occlusal plane. In view of the fact that both young and adult patients are seeking orthodontic treatment, another objective was to assess dependency between bone availability, sex, and age.

2. Materials and Methods

The study was approved by the bioethics committee of Jagiellonian University (number 1072.6120.132.2020).

2.1. Materials

The study material was 100 consecutively selected cone beam computed tomography scans (CBCT) of the maxilla of patients of the University Dental Clinic in Krakow (50 men and women each), taken for any reason between 2018 and 2021. Sample size determination showed that 91 patients would be suitable based on a margin of error of 0.05, confidence level 0.8, and population size of 200. Subjects included in the study met the following criteria: age of the patient >12 years, presence of upper first and second molar, absence of pathologies of the maxilla, and absence of clefts and maxillofacial syndromes. Patients with a history of orthodontic treatment, facial surgery, or facial trauma as well as images with artifacts were excluded from the research.

The CBCT scans were acquired with the OP 3D Pro (KaVo, Berlin, Germany). The protocol was: field of view 130 × 150 mm, average exposure time 8.5 s, average scanning time 39 s, average voxel size 380 μm^{-5} mA. For the analysis of the CBCT images, a medical diagnostic monitor (RadiForce MX215, EIZO, Viena, Austria) and InViVo Dental Viewer (Anatomage, Santa Clara, CA, USA) were used. All the scans were analyzed by a single trained and calibrated senior postgraduate trainee in orthodontics.

2.2. IZC Bone Thickness Measurements

IZC bone thickness was measured at the level of the mesiobuccal and distobuccal root of the first permanent molar and between the first and second molar as presented in the Figure 1. To localize the buccal roots of the first molar, a horizontal view was used. In the sagittal view, the horizontal axis was located at the level of $\frac{1}{2}$ height of the roots. Subsequently, a picture of the maxilla in horizontal projection was rotated in relation to the coronal axis to obtain an angle of 90° between the coronal axis and buccal surface of the alveolar process at the level of point M (mesial buccal root of the first molar), point D (distal buccal root of the first molar), and point I (interdental space between the first and the second molar) to imitate clinical conditions during mini-implant placement. Using the occlusal plane as a reference line, three points in the vertical plane were defined in modified coronal view at the levels of 12 mm, 14 mm, and 16 mm from the occlusal plane as points of potential mini-implants insertion. All the measurements were taken at an angle of 70 degrees to the occlusal plane, on the patients' left side.

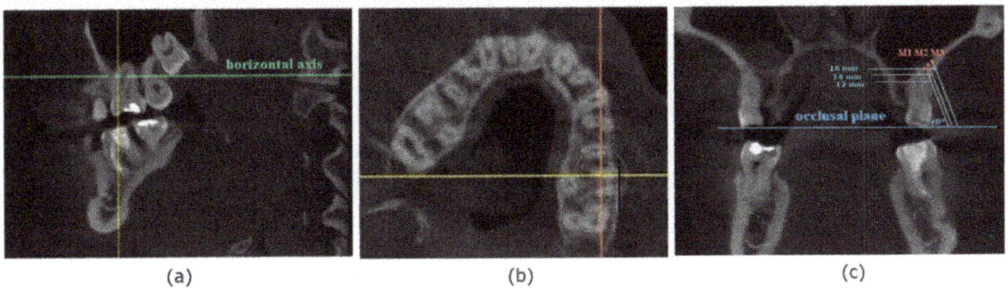

Figure 1. Measurement of bone depth on CBCT scans: (**a**)—sagittal slice consisting of mesiobuccal root of the first permanent molar used to obtain horizontal slice; (**b**)—horizontal view of the maxilla rotated in relation to the coronal axis to obtain a cross-section of the alveolar process at the level of M1; (**c**)—measurement of IZC thickness in modified coronal view.

2.3. Palatal Bone Thickness Measurements

Modified palatal height index was also calculated as a relation between the height and width of the palate according to the formula:

$$\text{Modified palatal height index} = \frac{\text{Palatal height} \times 100\%}{\text{Palatal width}}$$

For the purpose of this measurement, coronal slice of CBCT was selected, when the coronal axis was set at the center of the crown of the first molar. Although to calculate palatal height index [13] the distance between palatal cusps of the first maxillary molars should be used, for the purpose of our study, palatal width was defined as the horizontal distance between alveolar ridges on the palatal side of the right and left first maxillary molar to eliminate the influence of the inclination of these on the result (Figure 2). The perpendicular distance between the midpalatal raphe and the occlusal plane was measured to evaluate palatal height.

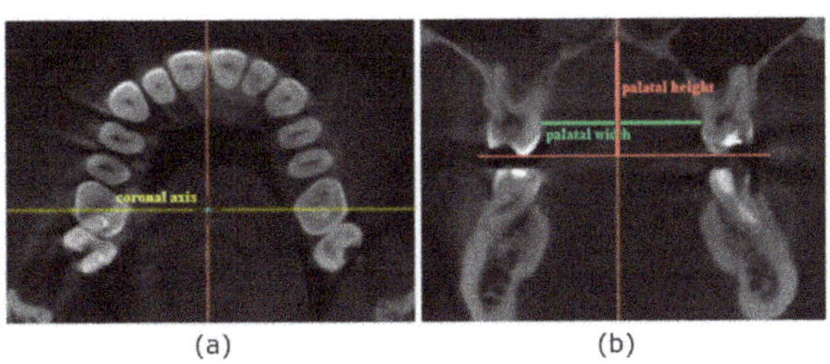

Figure 2. Measurement of modified palatal height index: (**a**)—horizontal slice used to obtain coronal view, (**b**)—measurements of the palatal width and height.

2.4. Intra-Examiner Error Calculation

After 3 weeks, measurements of 10 randomly selected CBCTs were repeated to calculate intra-examiner error and determine method reliability.

2.5. Statistical Analysis

Data analysis was performed using R software version 4.3.0 (R: A language and environment for statistical computing. R Foundation for Statistical Computing, Vienna, Austria).

To test whether the sample fits normal distribution, the Shapiro–Wilk test was used. Analysis of quantitative variables was performed by calculating mean, standard deviation, median, and quartiles. Qualitative variables were analyzed by calculating the number and percentage of occurrences of each value. The Mann–Whitney test was used to compare quantitative variables between two groups, while the Kruskal–Wallis test (followed by Dunn post-hoc test) was used for more than two groups. Concordance of measurements of quantitative variables was assessed with ICC (Intraclass Correlation Coefficient) type 2 (according to the Shrout and Fleiss classification). The Friedman test (followed by paired Wilcoxon tests with Bonferroni correction as a post-hoc procedure) was used to compare more than two repeated measures of quantitative variables. The relationship between two quantitative variables was assessed with Spearman's coefficient of correlation.

The significance level for all statistical tests was set to 0.05.

3. Results

3.1. Patients Characteristics

A total of 100 consecutively selected CBCT images of the maxilla (50 men and women each) that met the inclusion criteria were evaluated in this study. The mean ages of women and men were 28.7 and 28.92, respectively. Patients' age details are presented in Table 1.

Table 1. Patients characteristics.

Sex	N	Age							p
		Mean	SD	Median	Min	Max	Q1	Q3	
Female	50	28.70	13.83	25	12	65	16.5	38.75	$p = 0.654$
Male	50	28.92	11.95	27	13	63	18.0	36.75	
Total	100	28.81	12.86	25	12	65	18.0	38.00	

p—Mann–Whitney test, SD—standard deviation, Q1—lower quartile, Q3—upper quartile.

3.2. Intra-Examiner Error Calculation

Within a 3-week interval, 10 randomly selected CBCT scans were subjected to repeated measurements. The concordance of the measurements of quantitative variable assessment with intraclass correlation coefficient type 2 (according to the Shrout and Fleiss classification) indicated good or excellent accordance between the first and second measurements (Table 2).

Table 2. Intra-examiner error calculation.

Parameter	Measurement 1 (Mean ± SD)	Measurement 2 (Mean ± SD)	ICC	95% CI		Agreement (Cicchetti)	Agreement (Koo and Li)
M1	1.86 ± 2.29	1.88 ± 2.42	0.995	0.980	0.999	Excellent	Excellent
M2	2.84 ± 2.77	2.89 ± 2.84	0.995	0.983	0.999	Excellent	Excellent
M3	2.37 ± 2.13	2.79 ± 1.98	0.895	0.646	0.973	Excellent	Good
D1	4.67 ± 3.15	4.48 ± 3.25	0.987	0.950	0.997	Excellent	Excellent
D2	3.94 ± 2.29	3.87 ± 2.31	0.991	0.965	0.998	Excellent	Excellent
D3	2.91 ± 1.65	2.86 ± 1.61	0.959	0.853	0.989	Excellent	Excellent
I1	6.14 ± 1.34	6.01 ± 1.5	0.695	0.189	0.913	Good	Fair
I2	5.24 ± 1.91	5.47 ± 2.14	0.911	0.700	0.977	Excellent	Excellent
I3	4.52 ± 1.74	4.66 ± 1.7	0.974	0.904	0.993	Excellent	Excellent
Height of the palate	20.4 ± 2.53	20.42 ± 2.82	0.973	0.903	0.993	Excellent	Excellent
Width of the palate	33.91 ± 2.62	33.72 ± 2.75	0.965	0.873	0.991	Excellent	Excellent
Palatal height index	60.15 ± 5.83	60.53 ± 6.79	0.960	0.856	0.990	Excellent	Excellent

3.3. Bone Thickness

Tables 3–5 present differences in the mean bone thickness at the level of 12 mm, 14 mm, and 16 mm depending on the sagittal location of the mini-implant. At the height of 12 mm and 14 mm, bone layer was the thickest in the interdental space between the first and the second molar (6.03 ± 2.64 mm and 4.74 ± 2.17 mm, respectively) followed by the area of the distal root of the first molar (3.71 ± 2.76 mm and 3.11 ± 2.35 mm). In the case of the measurements performed at the height of 16 mm, the difference was statistically insignificant.

Table 3. Mean bone thickness at the height of 12 mm from the occlusal plane.

Bone Thickness (mm)	M1	D1	I1	p
mean ± SD	2.5 ± 2.55	3.71 ± 2.76	6.03 ± 2.64	$p < 0.001$
median	2.13	3.53	5.94	
quartiles	0–3.9	1.53–5.23	4.25–7.65	I1 > D1 > M1

p—Friedman test + post-hoc analysis (Wilcoxon paired tests with Bonferroni correction).

Table 4. Mean bone thickness at the height of 14 mm from the occlusal plane.

Bone Thickness (mm)	M2	D2	I2	p
mean ± SD	2.54 ± 2.42	3.11 ± 2.35	4.74 ± 2.17	$p < 0.001$
median	2.36	3.29	4.84	
quartiles	0–4.02	0.86–4.64	3.04–6.17	I2 > D2 > M2

p—Friedman test + post-hoc analysis (Wilcoxon paired tests with Bonferroni correction).

Table 5. Mean bone thickness at the height of 16 mm from the occlusal plane.

Bone Thickness (mm)	M3	D3	I3	p
mean ± SD	2.42 ± 2.16	2.59 ± 2.08	3.46 ± 1.93	$p = 0.453$
median	2.16	2.29	3.54	
quartiles	0–3.95	1–3.65	1.83–4.7	

p—Friedman test.

Tables 6–8 and Figure 3 present differences in bone thickness depending on the vertical location of the mini-implant. In the case of the area of the mesial root of the first molar, there was no significant difference in the bone thickness at different levels. In the case of the area of distal root and interdental space, the greatest values of bone thickness were recorded at the level of 12 mm from the occlusal plane (3.71 ± 2.76 mm and 6.03 ± 2.64 mm, respectively) and with the growing distance from the occlusal plane, the bone thickness presented a decreasing trend.

Table 6. Mean bone thickness at the level of the mesial root of the first molar.

Bone Thickness (mm)	M1	M2	M3	p
mean ± SD	2.5 ± 2.55	2.54 ± 2.42	2.42 ± 2.16	$p = 0.098$
median	2.13	2.36	2.16	
quartiles	0–3.9	0–4.02	0–3.95	

p—Friedman test.

Table 7. Mean bone thickness at the level of the distal root of the first molar.

Bone Thickness (mm)	D1	D2	D3	p
mean ± SD	3.71 ± 2.76	3.11 ± 2.35	2.59 ± 2.08	$p < 0.001$
median	3.53	3.29	2.29	
quartiles	1.53–5.23	0.86–4.64	1–3.65	D1 > D2 > D3

p—Friedman test + post-hoc analysis (Wilcoxon paired tests with Bonferroni correction).

Table 8. Mean bone thickness at the level of interdental space between the first and second molar.

Bone Thickness (mm)	I1	I2	I3	p
mean ± SD	6.03 ± 2.64	4.74 ± 2.17	3.46 ± 1.93	$p < 0.001$
median	5.94	4.84	3.54	
quartiles	4.25–7.65	3.04–6.17	1.83–4.7	I1 > I2 > I3

p—Friedman test + post-hoc analysis (Wilcoxon paired tests with Bonferroni correction).

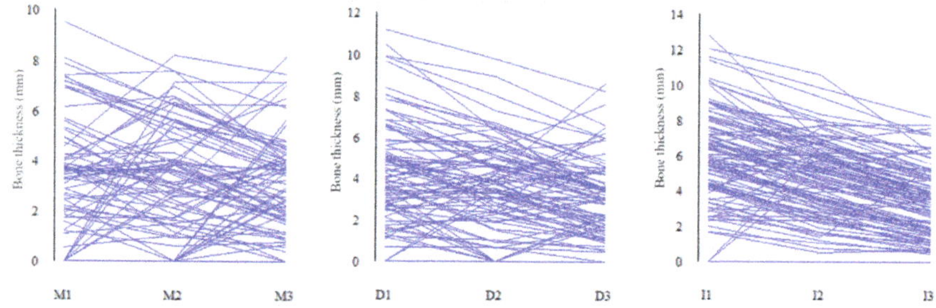

Figure 3. A parallel coordinate plot presenting bone thickness at the level of the mesial root of the first molar (M1, M2, M3), the distal root of the first molar (D1, D2, D3), and the interdental space between the first and the second molar (I1, I2, I3).

3.4. Correlation with Age

According to Spearman's correlation coefficient test, there was a significant and negative correlation between the bone thickness and the age in the case of measurement points M1, M2, M3, D1, D2, and D3 (Table 9). In points I1, I2, and I3 the correlation was also negative, but statistically insignificant. Results are also presented in Figure 4.

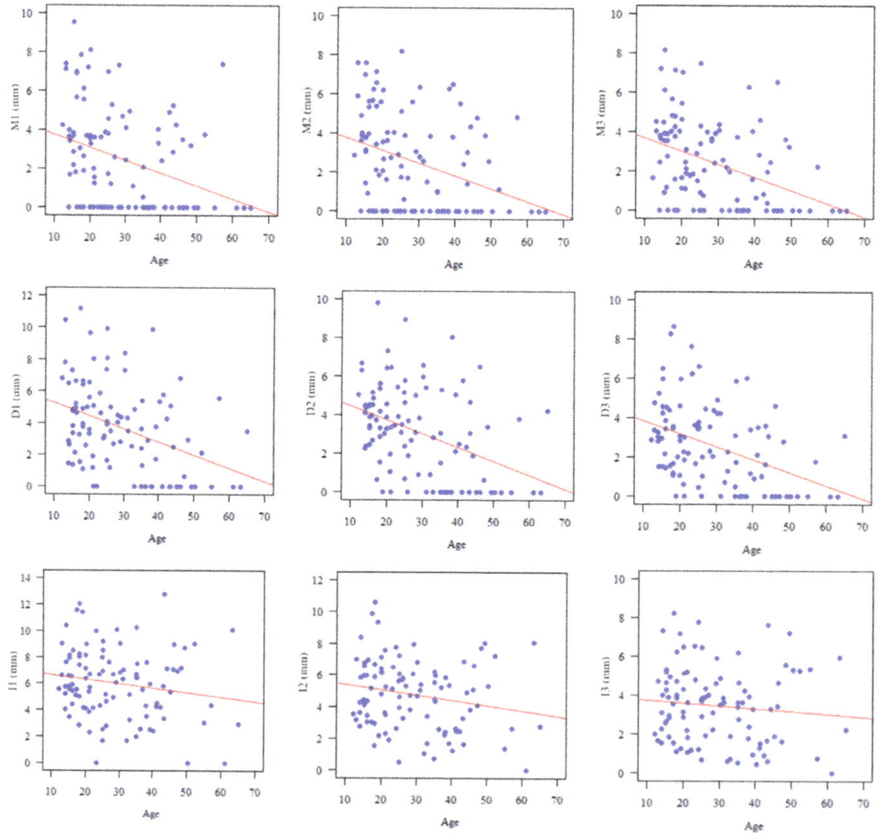

Figure 4. Correlation between the age of the patients and the bone thickness.

Table 9. Spearman's correlation coefficient test for comparison of parameters measured according to the age of the patients.

Parameter	Age
	Spearman's Correlation Coefficient
M1 (mm)	r = −0.37, $p < 0.001$ *
M2 (mm)	r = −0.381, $p < 0.001$ *
M3 (mm)	r = −0.424, $p < 0.001$ *
D1 (mm)	r = −0.378, $p < 0.001$ *
D2 (mm)	r = −0.406, $p < 0.001$ *
D3 (mm)	r = −0.42, $p < 0.001$ *
I1 (mm)	r = −0.147, $p = 0.144$
I2 (mm)	r = −0.178, $p = 0.077$
I3 (mm)	r = −0.107, $p = 0.289$

* statistically significant ($p < 0.05$).

3.5. Correlation with Sex

Table 10 and Figure 5 report the correlation between the bone thickness and sex of the patients. The only statistically significant dependency was found at the I2 and I3 points, where bone thickness was greater in male patients.

Table 10. Correlation between the bone thickness and sex of the patients.

Bone Thickness (mm)	Sex	N	Mean	SD	Median	Min	Max	Q1	Q3	p
M1	Female	50	2.96	2.77	3.29	0.00	9.53	0.00	4.86	$p = 0.078$
	Male	50	2.03	2.25	1.73	0.00	7.86	0.00	3.60	
M2	Female	50	2.65	2.16	2.71	0.00	7.59	0.15	4.00	$p = 0.438$
	Male	50	2.42	2.68	1.48	0.00	8.21	0.00	4.03	
M3	Female	50	2.32	1.96	1.96	0.00	7.20	0.79	3.78	$p = 0.794$
	Male	50	2.52	2.36	2.42	0.00	8.15	0.00	4.06	
D1	Female	50	3.54	2.64	3.50	0.00	9.87	1.40	5.05	$p = 0.616$
	Male	50	3.88	2.90	3.74	0.00	11.18	1.97	5.36	
D2	Female	50	3.00	2.12	3.04	0.00	8.04	1.62	4.20	$p = 0.565$
	Male	50	3.22	2.57	3.42	0.00	9.79	0.66	5.03	
D3	Female	50	2.44	1.83	2.14	0.00	6.51	1.06	3.50	$p = 0.691$
	Male	50	2.74	2.31	2.82	0.00	8.64	0.78	4.06	
I1	Female	50	5.53	2.40	5.61	0.00	10.10	4.18	7.52	$p = 0.11$
	Male	50	6.52	2.80	6.50	0.00	12.80	4.95	8.02	
I2	Female	50	4.14	2.09	3.94	0.00	8.01	2.51	5.91	$p = 0.012$ *
	Male	50	5.34	2.10	5.17	1.05	10.60	3.98	6.69	
I3	Female	50	2.86	1.74	2.84	0.00	7.21	1.26	3.99	$p = 0.003$ *
	Male	50	4.06	1.94	3.86	0.48	8.22	2.58	5.28	

p—Mann–Whitney test, SD—standard deviation, Q1—lower quartile, Q3—upper quartile. * statistically significant ($p < 0.05$).

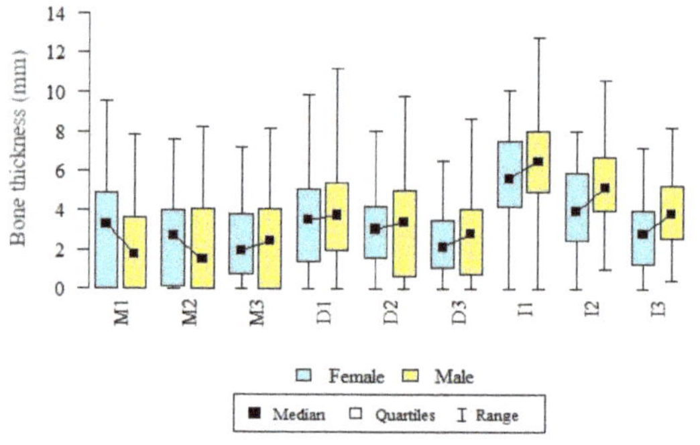

Figure 5. Graphic presentation of the correlation between the bone thickness and sex of the patients.

3.6. Correlation with Palatal Height Index

A significant and negative correlation between bone thickness and modified palatal height index was found in the case of points M1, D1, and D2 (Table 11, Figure 6). Nevertheless, it can be observed that bone depth tends to decrease with the increase of modified palatal index when areas buccal to the first molars are analyzed, and increase when interdental area is taken into consideration.

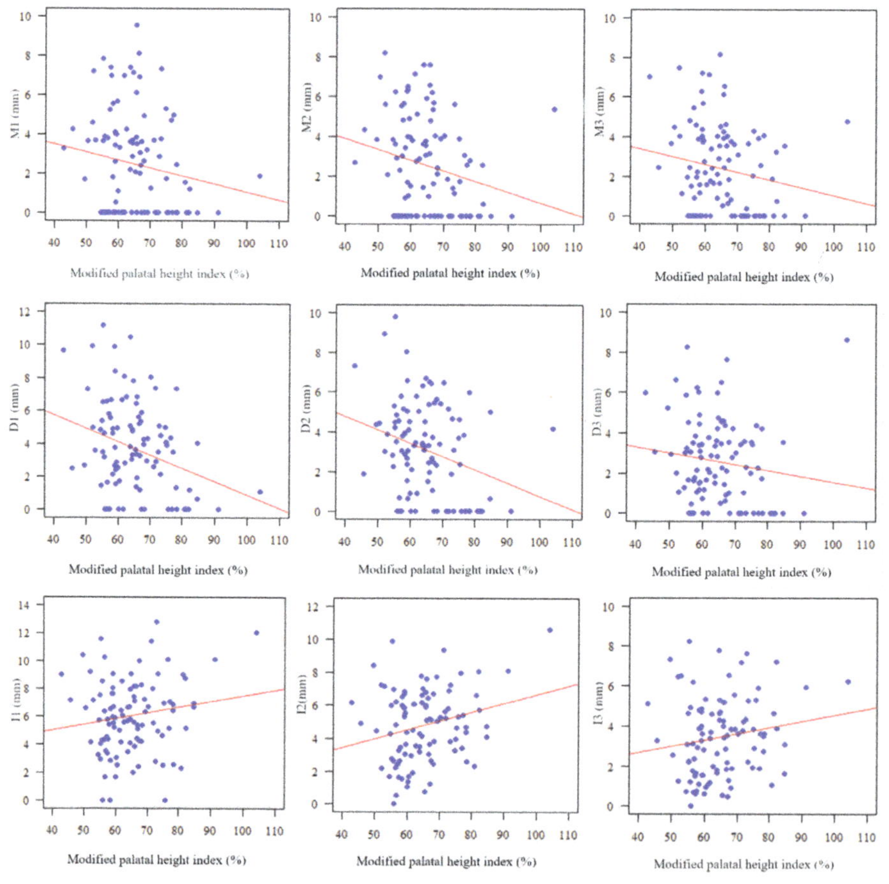

Figure 6. Correlation between the bone thickness and modified palatal height index.

Table 11. Correlation between the bone thickness and palatal height index.

Parameter	Modified Palatal Height Index
	Spearman's Correlation Coefficient
M1 (mm)	r = −0.113, p = 0.264
M2 (mm)	r = −0.206, p = 0.04 *
M3 (mm)	r = −0.151, p = 0.134
D1 (mm)	r = −0.208, p = 0.038 *
D2 (mm)	r = −0.249, p = 0.012 *
D3 (mm)	r = −0.184, p = 0.066
I1 (mm)	r = 0.091, p = 0.369
I2 (mm)	r = 0.175, p = 0.081
I3 (mm)	r = 0.147, p = 0.146

* statistically significant ($p < 0.05$).

4. Discussion

The area of the IZC is considered a safe zone for mini-screw placement, as this region is devoid of nerves, major blood vessels, or teeth [9]. Moreover, adequate bone density provides good primary stability, making IZC a valuable source of anchorage. Nevertheless, the close proximity of the maxillary sinus and the risk of Schneiderian membrane perforation, whose integration is vital for undisturbed sinus function [14], involves the need to accomplish the implantation procedure in a precise manner. As stated in the literature, when an implant invades the maxillary sinus less than 2 mm, the Schneiderian membrane is elevated and new bone is formed in this region. In the case of deeper insertion, the maxillary membrane becomes perforated, which may promote the development of sinusitis [15].

Theoretically, to obtain sufficient stability of the IZC mini-screw a minimum bone thickness of 6 mm is required [7]. Liou et al. report that the mean bone depth measured in Taiwan patients at the level of the mesial root of the first maxillary molar, 14 mm above the occlusal plane, at the insertion angle of 70° was 8 mm. According to our study, mean bone thickness at the same point and angle of insertion was 2.54 ± 2.42 mm, which implies much poorer conditions for implantation. Matias et al. conducted a study on Brazilian patients to identify the optimal areas for extra-alveolar mini-screws of patients with different facial patterns [16]. There was no significant difference in the case of IZC bone thickness among groups, but the mean bone depth at the level of the distal root of the first maxillary molar 13 mm above the occlusal plane and with the insertion angle of 70° ranged from 7.11 ± 1.95 mm in dolichofacial patients to 7.51 ± 2.16 mm in brachyfacial patients, which gives much higher results when compared to analogical measurements performed in our study (3.71 ± 2.76 mm at the height of 12 mm and 3.11 ± 2.35 mm at the height of 14 mm). Values obtained at the level of interproximal contact of the first and second maxillary molars in another Brazilian study (7.3 ± 3.0 mm) also surpass our results (6.03 ± 2.64 mm) [17]. Nevertheless, to confirm that the mean bone thickness of the IZC in Caucasian patients is lower when compared to the non-Caucasian population, multicenter studies in a much larger group of patients are needed.

The most favorable conditions for IZC mini-screws placement were found at the level of interdental space between the first and the second molar, where bone depth ranged from 3.46 ± 1.93 mm at the point I3 to 6.03 ± 2.64 mm at I1. The thinnest bone was identified at the area of the mesial root of the first molar (from 2.42 mm to 2.54 mm), which indicates that bone thickness increases from front to back. A study conducted by Amri et al. [18] on Arabian patients, according to which mean bone depth at the level of the mesial root of the first molar with an insertion angle of 70° was 3.90 mm, confirmed our conclusion that the available bone in this area is insufficient and the risk of sinus injury could be high.

According to Matias et al., mean bone depth tends to decrease with an increase in the distance from the occlusal plane (an increase of 2 mm in height was connected with a 2 mm decrease in the bone thickness) [16]. Although the results of our study seem to be in line with the abovementioned theory, the decrease in bone depth with growing vertical distance from the occlusal plane is not so notable. The mean difference in bone depth between the insertion point at the height of 12 mm and 16 mm is 0.8 mm in the case of the area of the mesial root of the first molar, 1.12 mm at the area of the distal root of the first molar, and 2.57 mm in the interdental space.

Results of our study indicate the presence of a significant and negative correlation between the age of the patients and bone thickness buccally to the first molar. Similar correspondence can be also observed in the case of interdental point of insertion, but the results are not statistically significant. Only a few studies assessed age dependency. Amri et al. [18] did not notice age dependency in the case of insertion site between the first and the second molar among patients aged 18 to 42 years.

Considering differences in bone thickness among sexes, bigger values were observed in the case of male patients, but the only significant differences were found in the I2 and I3 measurements. Similarly, Santos et al. [19] and Amri et al. [18] did not find any differences in the bone thickness buccally to the distal and mesial root of the maxillary first molar respectively, comparing sexes.

Since there are studies where some correlation between facial pattern and bone availability at the IZC area has been found [20], the aim of our study was also to observe dependency between the shape of the maxilla, expressed as a modified palatal index, and the structure of the IZC region. Although it needs to be emphasized that significant results were obtained only in the case of M2, D1, and D2 points, the bone thickness buccally to the first molar tends to decrease with the increase of modified palatal height index. Interestingly, the bone depth In the interdental space tends to grow slightly with the increase of the modified palatal height index. To conclude, a narrow or high maxilla might be associated with a thin buccal bone plate in the area of the first molar and superior implantation conditions between the first and second molar. A study by Tavares et al. partially supports our results [17]. They found that bone thickness between the first and the second molar was greater in dolichofacial patients when compared to meso- and brachyfacial individuals. On the other hand, Matias et al., in a similar study, did not observe significant differences in the IZC thickness among groups [16]. Inconsistencies in the above-mentioned investigations may result from the different methods of classification of patients to brachy-meso- and dolichofacial groups (based on the angle between SN and Go-Gn line or VERT index). It should be also accentuated that the above-mentioned measurements might be strongly affected by the morphology of the mandible. Consequently, the palatal index used in our study may provide more uninfluenced information regarding spatial conditions for the implantation of mini-screws.

Considering the practical implications of the abovementioned research, it can be stated that in Polish patients, the mean bone depth at the area of infrazygomatic ridge rarely reaches a minimum recommended value of 6 mm. Surprisingly, our clinical experience indicates a relatively good incidence of IZC mini-screws survival, even in the absence of bone of sufficient thickness (Figure 7 presents a CBCT scan of IZC mini-screw application in the case of small bone depth).

Some authors attribute this high success rate to the penetration through two cortical layers, which guarantees more than 1 mm of the cortical bone and improves the primary stability of mini-implants [5,7]. Nonetheless, insufficient bone depth might be attributed to the increased risk of maxillary sinus membrane rupture. As stated in the literature, sinus invasion during IZC mini-screws placement is relatively common. Although according to the Chinese study, 78% of mini-implants penetrated into the maxillary sinus [5], irritation of the Schneiderian membrane measured as the increase in its thickness was significant only when mini-implants penetrated into the sinus deeper than 1 mm. Minor, uncomplicated injuries of the maxillary sinus by miniscrews may regenerate spontaneously and orthodontic treatment interruption and mini-implant removal are not recommended [8,21].

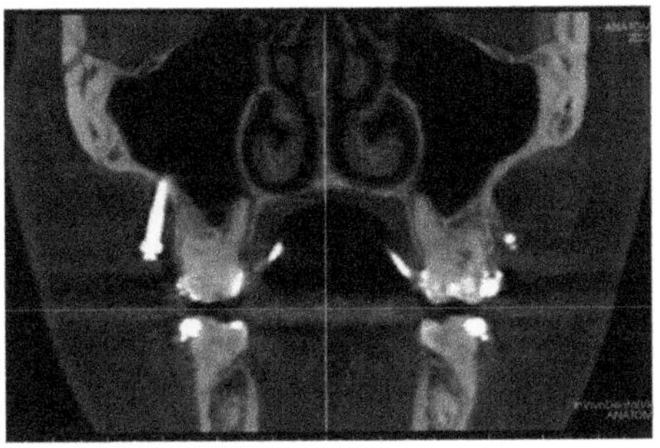

Figure 7. IZC mini-screw with an insertion depth of 1.5 mm, successfully used for the dystalization in the upper arch.

Although the findings of this study shed some light on IZC mini-screw placement, they have to be seen in light of certain limitations. One of them is the limited sample size coming from one center, so the results of the present research should not be extrapolated to the entire Polish population. Another limitation concerns the 380 μm voxel size—a smaller voxel size might provide even more precise measurements.

5. Conclusions

Considering sagittal and vertical dimensions, the most satisfactory conditions for IZC mini-screws placement were localized at the level of interdental space between the first and second molar, 12 mm above the occlusal plane, where mean bone depth reached 6.03 ± 2.64 mm. The thinnest bone depth was localized at the level of the mesial root of the first molar, 16 mm above the occlusal plane (2.42 ± 2.16). There was a significant and negative correlation between bone thickness and age in the case of measurements taken buccally to the first molar. Due to the fact that individual variation in the growth and development of the maxilla and maxillary sinus may affect the anatomy of the IZC area, CBCT scan analysis should be considered prior to the mini-screw implantation procedure.

Author Contributions: Conceptualization, M.G.-S.; methodology, M.G.-S.; software, M.G.-S., J.Ś., M.U. and S.Ż.; validation, M.P.; formal analysis, M.G.-S., J.Ś., M.U., S.Ż. and M.P.; investigation, M.G.-S., J.Ś. and M.U.; resources, M.G.-S. and M.P.; data curation, M.G.-S.; writing—original draft preparation, M.G.-S.; writing—review and editing, J.Ś., M.U., S.Ż. and M.P.; visualization, M.G.-S.; supervision, M.P.; project administration, M.G.-S.; funding acquisition, M.G.-S. and M.P. All authors have read and agreed to the published version of the manuscript.

Funding: This research received no external funding.

Institutional Review Board Statement: The study was conducted according to the guidelines of the Declaration of Helsinki and approved by the bioethics committee of Jagiellonian University (number 1072.6120.132.2020).

Informed Consent Statement: Not applicable.

Data Availability Statement: The datasets generated and/or analyzed during the current study are available from the corresponding author upon reasonable request.

Conflicts of Interest: The authors declare no conflict of interest.

References

1. De Clerck, H.; Geerinckx, V.; Siciliano, S. The Zygoma Anchorage System. *J. Clin. Orthod. JCO* **2002**, *36*, 455–459. [PubMed]
2. Chung, K.-R.; Kim, S.-H.; Kang, Y.-G.; Nelson, G. Orthodontic miniplate with tube as an efficient tool for borderline cases. *Am. J. Orthod. Dentofac. Orthop.* **2011**, *139*, 551–562. [CrossRef]
3. Chang, C.C.H.; Lin, J.S.Y.; Yeh, H.Y. Extra-Alveolar Bone Screws for Conservative Correction of Severe Malocclusion Without Extractions or Orthognathic Surgery. *Curr. Osteoporos. Rep.* **2018**, *16*, 387–394. [CrossRef] [PubMed]

4. Uribe, F.; Mehr, R.; Mathur, A.; Janakiraman, N.; Allareddy, V. Failure rates of mini-implants placed in the infrazygomatic region. *Prog. Orthod.* **2015**, *16*, 31. [CrossRef]
5. Jia, X.; Chen, X.; Huang, X. Influence of orthodontic mini-implant penetration of the maxillary sinus in the infrazygomatic crest region. *Am. J. Orthod. Dentofac. Orthop.* **2018**, *153*, 656–661. [CrossRef]
6. Lovely; Pradeep, R.; Reddy, C.M. Clinical applicability of IZC's in orthodontics—A review. *J. Contemp. Orthod.* **2023**, *6*, 172–177. [CrossRef]
7. Liou, E.J.W.; Chen, P.H.; Wang, Y.C.; Lin, J.C.Y. A computed tomographic image study on the thickness of the infrazygomatic crest of the maxilla and its clinical implications for miniscrew insertion. *Am. J. Orthod. Dentofac. Orthop.* **2007**, *131*, 352–356. [CrossRef] [PubMed]
8. Motoyoshi, M.; Yoshida, T.; Ono, A.; Shimizu, N. Effect of cortical bone thickness and implant placement torque on stability of orthodontic mini-implants. *Int. J. Oral Maxillofac. Implant.* **2007**, *22*, 779–784.
9. Sandeep, S.; Katheesa, P. An overview of extra alveolar bone screws-IZC/BS screws. *Biomedicine* **2020**, *40*, 108–110. [CrossRef]
10. Lin, J.J. A New method of placing orthodontic bone screws in IZC. *A News Trends Orthod.* **2009**, *13*, 4–7.
11. Lin, J.; Eugene, R.W. CBCT imaging to diagnose and correct the failure of maxillary arch retraction with IZC screw anchorage. *Int. I Orthop. Implantol.* **2014**, *35*, 4–17.
12. Ghosh, A. Infra-Zygomatic Crest and Buccal Shelf-Orthodontic Bone Screws: A Leap Ahead of Micro-Implants-Clinical Perspectives. *J. Indian Orthod. Soc.* **2018**, *52*, 127–141. [CrossRef]
13. Aluru, Y.; Rng, R.; Gujar, A.N.; Kondody, R. Correlation of Palatal Index with Pharyngeal Airway in Various Skeletal Patterns. *Cureus* **2023**, *15*, e39032. [CrossRef] [PubMed]
14. Ardekian, L.; Oved-Peleg, E.; Mactei, E.E.; Peled, M. The clinical significance of sinus membrane perforation during augmentation of the maxillary sinus. *J. Oral Maxillofac. Surg.* **2006**, *64*, 277–282. [CrossRef] [PubMed]
15. Reiser, G.M.; Rabinovitz, Z.; Bruno, J.; Damoulis, P.D.; Griffin, T.J. Evaluation of maxillary sinus membrane response following elevation with the crestal osteotome technique in human cadavers. *Int. J. Oral Maxillofac. Implant.* **2002**, *16*, 833–840.
16. Matias, M.; Flores-Mir, C.; de Almeida, M.R.; Vieira, B.d.S.; de Freitas, K.M.S.; Nunes, D.C.; Ferreira, M.C.; Ursi, W. Miniscrew insertion sites of infrazygomatic crest and mandibular buccal shelf in different vertical craniofacial patterns: A cone-beam computed tomography study. *Korean J. Orthod.* **2021**, *51*, 387–396. [CrossRef]
17. Tavares, A.; Montanha-Andrade, K.; Ramos Cury, P. Tomographic assessment of infrazygomatic crest bone depth for extra-alveolar miniscrew insertion in subjects with different vertical and sagittal skeletal patterns. *Orthod. Craniofacial Res.* **2021**, *25*, 49–54. [CrossRef]
18. Al Amri, M.S.; Sabban, H.M.; AlSaggaf, D.H.; Alsulaimani, F.F.; Al-Turki, G.A.; Al-Zahrani, M.S.; Zawawi, K.H. Anatomical consideration for optimal position of orthodontic miniscrews in the maxilla: A CBCT appraisal. *Ann. Saudi Med.* **2020**, *40*, 330–337. [CrossRef]
19. Santos, A.R.; Castellucci, M.; Crusoé-Rebello, I.M.; Sobral, M.C. Assessing bone thickness in the infrazygomatic crest area aiming the orthodontic miniplates positioning: A tomographic study. *Dent. Press J. Orthod.* **2017**, *22*, 70–76. [CrossRef] [PubMed]
20. Ozdemir, F.; Tozlu, M.; Germec-Cakan, D. Cortical bone thickness of the alveolar process measured with cone-beam computed tomography in patients with different facial types. *Am. J. Orthod. Dentofac. Orthop.* **2013**, *143*, 190–196. [CrossRef]
21. Baumgaertel, S.; Hans, M.G. Assessment of infrazygomatic bone depth for mini-screw insertion. *Clin. Oral Implant. Res.* **2009**, *20*, 638–642. [CrossRef] [PubMed]

applied sciences

MDPI

Article

A Digital 3D Retrospective Study Evaluating the Efficacy of Root Control during Orthodontic Treatment with Clear Aligners

Monica Macrì *, Silvia Medori, Giuseppe Varvara and Felice Festa

Department of Innovative Technologies in Medicine & Dentistry, University "G. d'Annunzio" of Chieti-Pescara, 66100 Chieti, Italy
* Correspondence: m.macri@unich.it

Abstract: This study aimed to investigate the efficacy of torque movement and the incidence of root resorption in the maxillary and mandibular teeth with clear aligner therapy using cone-beam computed tomography. The sagittal root positions, the faciolingual inclinations, and the root lengths of 672 teeth, from central incisors to first molars for each arch, were measured and compared on virtual cross sections from pre-treatment and post-treatment cone-beam computed tomography of 28 patients who received comprehensive orthodontic treatment with clear aligners. An improvement of root position was found in incisors, canines, and premolars of the upper and lower arches: over 78% of their root was centered in the alveolus at the end of orthodontic treatment. There was a statistically significant torque increase for incisors, canines, and first premolars at the end of therapy. The most considerable torque changes were achieved in incisors and canines, while the lowest was in posterior teeth. The maxillary and mandibular central incisors achieved $3.26 \pm 1.95°$ and $2.97 \pm 2.53°$ of mean torque increase, respectively. The root length loss was greater in the upper and lower central incisors. All teeth showed mild resorption (<10%) except for two upper lateral incisors, which showed moderate resorption (10.79% and 10.23%). Comprehensive treatment with clear aligners improved sagittal root position and increased torque, especially in the anterior teeth. Most teeth showed mild resorption after clear aligner therapy, and only two showed moderate resorption.

Keywords: 3D; artificial intelligence and health; CBCT; digital health; emerging technologies

Citation: Macrì, M.; Medori, S.; Varvara, G.; Festa, F. A Digital 3D Retrospective Study Evaluating the Efficacy of Root Control during Orthodontic Treatment with Clear Aligners. *Appl. Sci.* **2023**, *13*, 1540. https://doi.org/10.3390/app13031540

Academic Editors: Giuseppe Minervini and Rocco Franco

Received: 9 December 2022
Revised: 21 January 2023
Accepted: 22 January 2023
Published: 25 January 2023

1. Introduction

Clear aligner therapy, consisting of customized, removable appliances, has been widely used in clinical practice as a more aesthetic and comfortable alternative to multibracket appliances.

In the beginning, the aligners were limited only to mild malocclusions, such as anterior crowding, or to periodontal patients; through the years, thanks to advances in technology and clinical trials, clear aligners have effectively performed major tooth movements, e.g. premolar derotation as well as molar distalization [1]. Despite the predictability of the treatment, its clinical potency remains debatable; opponents have remarked on the need to require mid-course correction or case refinement, especially when treating complex malocclusions, whereas advocates have remained convinced of successful outcomes at the end of the therapy [2].

Compared with our early ancestors, the modern human face reveals a characteristic spatial distribution of bone deposition and resorption. In humans, the anterior portions of the maxilla and mandible's sub-nasal region are more susceptible to surface resorption during development [3,4]. Furthermore, in the sagittal projection of X-ray examinations, clinicians commonly find that the roots of teeth, especially in the anterior region, are positioned against the labial cortical plate. Therefore, it is fundamental to manage the radicular torque and the root position relative to the orofacial cortical plates during orthodontic treatment.

In fixed orthodontic therapy, torque expression depends on several factors: bracket prescription and material (metal or ceramic brackets), inter bracket distance, the vertical position of the bracket, tooth morphology, and mode of ligation, as well as size and quality of the wire [5]. Studies have reported that slot-arch wire interaction may not produce the 3D control required to express acceptable 3rd-order movement [6,7]. On the contrary, aligners offer customized prescriptions with none of the disadvantages related to bracket design and positioning or tooth morphology, thanks to peculiar features in the cervical region of the clear aligner, such as power ridges.

One of the side effects that can occur during orthodontic treatment is external apical root resorption (EARR). There are few studies on EARR with clear aligners, and their conclusions still need to be made more public [8]. Most studies suggested that the incidence and severity of EARR with clear aligners were lower than those with fixed appliances [9,10], whereas other research [11] found no significant differences between them. Intermittent force, light force, and shorter treatment duration may be reasons for the minimal EARR with clear aligners [8].

Most previous studies used panoramic or periapical radiographs to evaluate EARR, which may cause distortion and overestimate or underestimate the extent of resorption. In recent studies, cone-beam computed tomography (CBCT) has overcome these shortcomings and improved the accuracy of measuring root length [12,13]. CBCT data are highly reproducible [14] and offer excellent sensitivity and specificity [15].

Consequently, this retrospective study aimed to evaluate, by CBCT, the efficacy of torque movement on the upper and lower teeth during clear aligner treatment, and to investigate the incidence and severity of EARR.

2. Materials and Methods

The study group included 672 teeth from 28 subjects (17 females and 11 males) treated with clear aligners between April 2017 and January 2022 in the Department of Innovative Technologies in Medicine & Dentistry at "G. d'Annunzio" University of Chieti-Pescara. Ethics approval (number 23) was obtained by the hospital's Independent Ethics Committee of Chieti. The study protocol was drawn following the European Union Good Practice Rules, and the Helsinki Declaration.

The patients' ages are between 18 and 38 years, with an average of 27 years and 6 months. The treatment did not require extractions but did require the dentoalveolar expansion of the maxillary arch to resolve crowding and allow spontaneous mandibular advancement. All subjects requested more than 14 aligners for the arch (comprehensive treatment). Each aligner was changed every 14 days and was worn at least 22 h/day; the average duration of treatment was 25 months.

All subjects met the following inclusion criteria: (1) growth was completed; (2) sufficient height of clinical crown; (3) non-extraction therapy in which the presence of crowding can be managed with protrusion/proclination, expansion, and inter-proximal reduction (IPR); (4) good compliance during treatment; and (5) anti-tightening therapy and physiotherapy with stretching and strengthening exercises of the paravertebral muscles before starting orthodontic therapy.

The exclusion criteria were as follows: (1) systematic disease or drug-taking history affecting tooth movement; (2) orofacial malformation syndromes; (3) periodontal disease; (4) missing teeth (except for third molars); (5) extraction cases; (6) auxiliary treatment during arch expansion stage such as crossbite elastics; (7) surgical case; and (8) previous orthodontic treatment.

At the first visit (T0), records for each participant were collected, consisting of the following: (1) general and dental anamnesis; (2) extraoral and intraoral orthodontic clinical examination; (3) gnathological clinical examination; and (4) visual analogue scale (VAS) and muscular palpation to estimate the pain intensity ratio on patient's face and neck [16].

Each patient underwent a CBCT scan using Planmeca Promax® 3D MID unit (Planmeca Oy, Helsinki, Finland) according to the low dose protocol with these parameters: acquisition time of 15 s, 80 kVp, 5 mA, 35 microSievert (μSv), the field of view (FOV) of 240 × 190 mm, and normal image resolution [17]. The patient's CBCT was performed with the head oriented according to the Natural Head Position (NHP); the patient was in a sitting position with the back perpendicular to the floor as much as possible. The head was stabilized with ear rods in the external auditory meatus. The patient was instructed to look into their eyes in a mirror 1.5 m in front of them to obtain NHP. The NHP is a physiological and reproducible posture defined for the morphological analysis described in the orthodontic and anthropological literature [18]. Each subject was informed about the radiographic procedure and required to avoid movement and keep centric occlusion with the lip in light contact.

After X-ray scanning, DICOM (Digital Imaging and Communications in Medicine) image files were processed by Dolphin Imaging 3D software (Dolphin Imaging & Management Solutions, Chatsworth, CA) for storage and interpretation. Establishing a predefined patient's head orientation is necessary to obtain a predictable and repeatable three-dimensional (3D) analysis. The skull image was oriented according to NHP in the three planes of space perpendicular to each other, as shown in previous studies [19]: the transverse plane coincides with the Frankfurt plane (FH), a plane passing through two points: Orbital (Or) and Porion (Po); the sagittal plane coincides with the mid-sagittal plane (MSP), a plane perpendicular to the FH plane and passing through two points: Crista Galli (Cg) and Basion (Ba); the coronal plane coincides with the anteroposterior (PO) plane, perpendicular to the FH and MSP, passing through the right and left portion.

After the orientation of the head, the virtual 2D radiograms were extracted in the following sequence:

Lateral teleradiography, on which the cephalometric analysis, according to McLaughlin, is performed

- orthopantomography,
- TMJ stratigraphy,
- cross sections,
- posteroanterior teleradiography,
- superior and inferior submento-vertex
- virtual reconstruction of right and left masseter muscles.

Subsequently, extraoral photos (patient's face in frontal, in the right, and left side views) and intraoral photos (frontal, right, and left lateral photos, and upper and lower occlusal photos) were performed, and the dental arches were scanned using an intraoral scanner, which allows detecting details with an accuracy up to 7 μm.

The virtual setup for each subject was planned, and aligners were manufactured.

At the end of clear aligner therapy (T1), extraoral and intraoral photos, pain assessments (through VAS and muscular palpation), and a CBCT scan were taken for each patient, and 2D virtual radiograms were obtained, as previously described.

For each upper and lower tooth, from right to left, the first molar, the changes in root position, torque, and root length were evaluated by analyzing the cross sections at the start (T0) and the end (T1) of the treatment.

Table 1 shows the number of measurements for the type of tooth of each arch taken into consideration in this study.

Table 1. Number of measurements for type of tooth in each arch.

Tooth	Measurements T0 (n°)	Measurements T1 (n°)	Tot. Measurements (n°)
Central incisor	56	56	112
Lateral incisor	56	56	112
Canine	56	56	112
First premolar	56	56	112
Second premolar	56	56	112
First molar	56	56	112
From central incisor to first molar	336	336	672

The effectiveness of movement, that is, the evaluation of the sagittal root position in the alveolar bone, was performed by comparing pre-treatment and post-treatment root positions relative to the orofacial cortical plates in the cross sections at T0 and T1 stages.

The sagittal root position was qualitatively evaluated in the midsagittal view according to the rating scale reported by Kan et al. [20] and modified by Aman et al. [21] (Figure 1): In class I, the root is positioned against the labial cortical plate (A); in Class II, the root is centred in the middle of the alveolar housing without engaging either the labial or the palatal cortical plates at the apical third of the root (B); Class III, the root is positioned against the palatal cortical plate (C); Class IV, at least two-thirds of the root is engaging both the labial and palatal cortical plates (D); and Class V, the root is positioned outside the labial cortical plate (E).

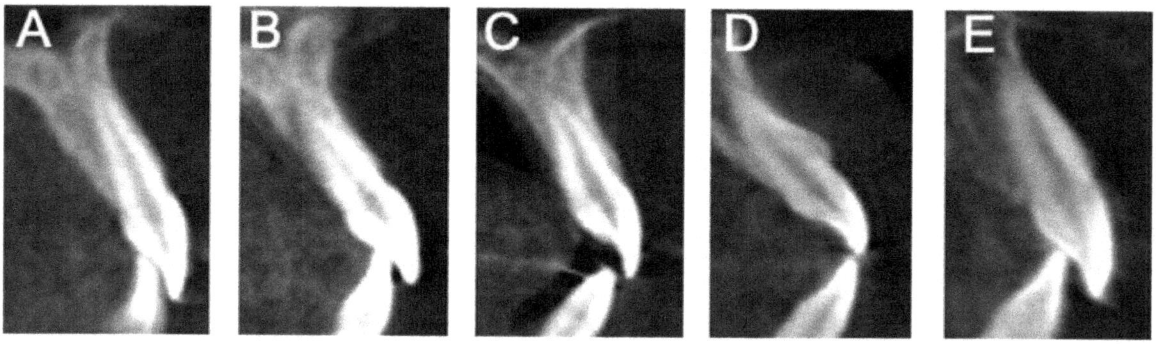

Figure 1. Classification of root position relative to cortical plates according to Aman: (**A**), Class I: the root is positioned against the labial cortical plate. (**B**), Class II: the root is centered in the alveolar housing without engaging the labial or palatal cortical plate at the apical third of the root. (**C**), Class III: the root is positioned against the palatal cortical plate. (**D**), Class IV: at least two-thirds of the root engages the labial and palatal cortical plates. (**E**), Class V: the root is positioned outside the labial cortical plate.

Pre-treatment and post-treatment faciolingual inclinations, that is, torque, of each upper and lower tooth from right to left first molar, were measured as the angle formed by the projection of the tooth's long axis on the faciolingual plane and the line of intersection between the faciolingual and mesiodistal planes by the University of Southern California (USC) root vector analysis software program [22] (Figure 2).

If the root center were lingual to the crown center, the torque measurement would be positive; otherwise, it would be negative. The custom USC root vector analysis program uses algorithms to calculate the torque values for all teeth automatically [22]. Subsequently, we reported the torque data in an Excel spreadsheet (version 2019; Microsoft, Redmond, Wash). The difference between pre-treatment and post-treatment torque values for each tooth and the percentage of torque variation were calculated using an Excel spreadsheet.

Additionally, we analyzed the presence of external apical root resorption (EARR). The root length was measured by Dolphin software as the perpendicular distance between the most apical point of the tooth and the reference line at the cementoenamel junction, according to the studies reported by Aman et al. [21] and Jiang et al. [2] (Figure 3).

EARR was defined as a root length loss (mm) between T0 and T1 stages, and the percentage of EARR was calculated as (root length loss/original root length) $\times$ 100% using an Excel spreadsheet. The severity of EARR was divided into the following three degrees according to the percentage of EARR: mild (<10%), moderate (10–20%), and severe (>20%).

Two blinded observers, previously instructed to use Dolphin Imaging 3D software, measured the torque and root length values. All measurements were repeated for 10% of the sample after 4 weeks.

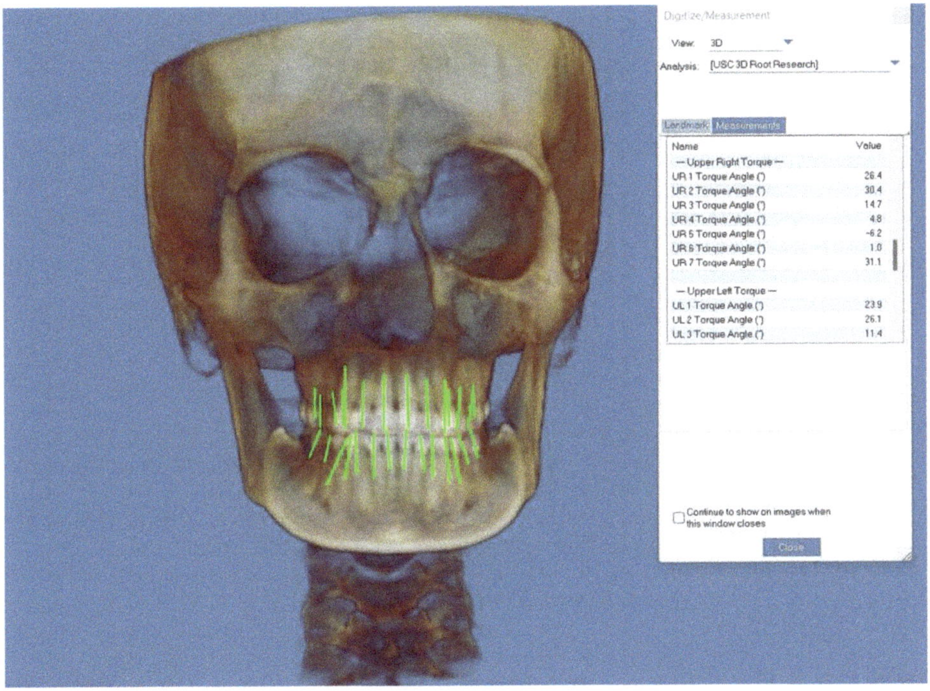

Figure 2. Measurements of torque using root vector analysis software program.

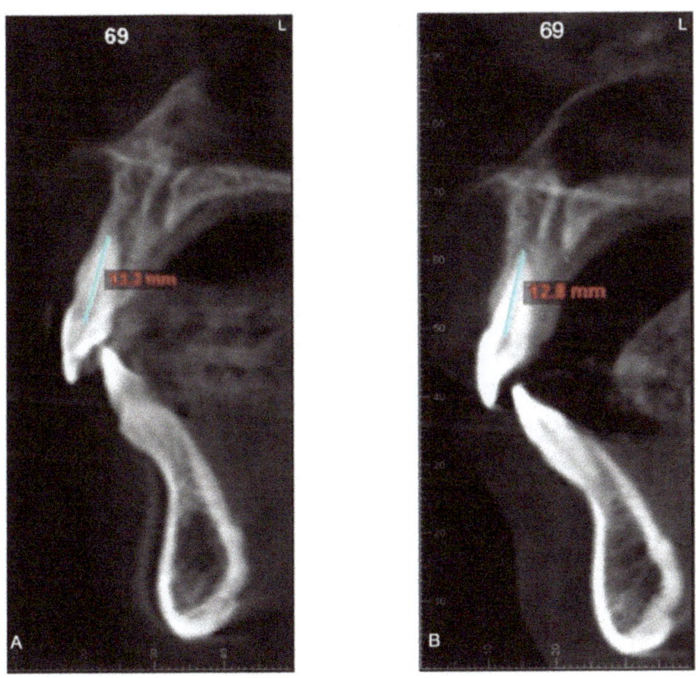

Figure 3. Measurements of the root length of the right central incisor on the cross sections: (**A**), pre-treatment; (**B**), post-treatment. This tooth shows a slight root resorption of 0.5 mm, equal to 3.7% of the initial root length.

The torque and root length values before and after therapy were subjected to statistical analysis to establish whether torque and root length changes are attributable to orthodontic therapy with clear aligners.

Intraclass correlation coefficients (ICCs) were used to estimate intra-rater and inter-rater reliabilities.

Statistical analyses were performed using Microsoft Excel (version 2019; Microsoft, Redmond, Wash) and StatPlus (AnalystSoft Inc., Walnut, CA, USA). A paired *t*-test was selected to compare torque and root length variations before and after therapy. The level of significance was set at 5%.

3. Results

The study analyzed 672 teeth of the upper and lower arches before and after therapy from 28 patients treated with clear aligners.

The ICC tests showed high intra-rater (0.9886 and 0.9883 for torque, 0.9839 and 0.9845 for root length) and inter-rater reliabilities (0.9884 for torque and 0.9880 for root length).

To evaluate the changes in sagittal root positions relative to the orofacial cortical plates, the positions of the roots at the T0 and T1 stages were compared on the cross sections using the classification proposed by Kan et al. [20] and modified by Aman et al. [21].

In the upper and lower arches, no roots were found positioned against the palatal cortical plate or both cortical plates at either T0 or T1 stages.

As regards the upper arch (Figure 4), an improvement in root position has been noticed, especially in the anterior sectors.

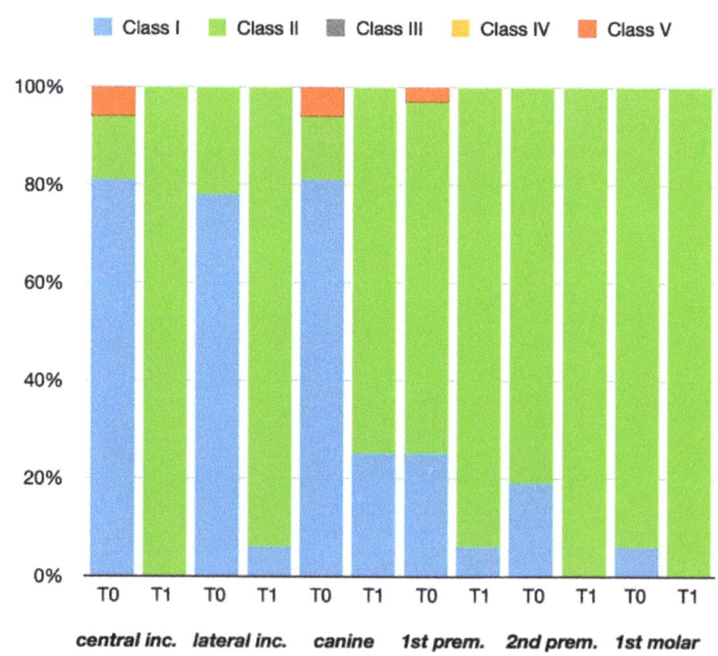

Figure 4. Percentage of sagittal root positions before and after therapy in the upper arch.

The central incisors achieved more significant progress. Indeed, all roots were correctly positioned at the end of treatment; the canines showed lower effectiveness since 22% of their roots remained against the labial cortical plate at T1. For the lateral incisors and first premolars, all roots were in Class II al T1, except for only 6% positioned against the buccal cortical plate.

In the second bicuspid, 19% of their roots in class I at T0 were centered in the alveolus at T1. In the first molars, 6% of roots in contact with the labial cortical plate at T0 were found in class II at T1, so 100% of their roots were in the center of the alveolus at the end of therapy.

The aligners satisfactorily managed the roots of the incisors, canines, and first premolars positioned outside the labial cortical plate at T0.

In the mandibular arch (Figure 5), the management of root position was better in the central incisors (from 22% in class II at T0 to 84% in class II at T1). However, 16% of the roots remained against the buccal cortical plate at the end of the therapy.

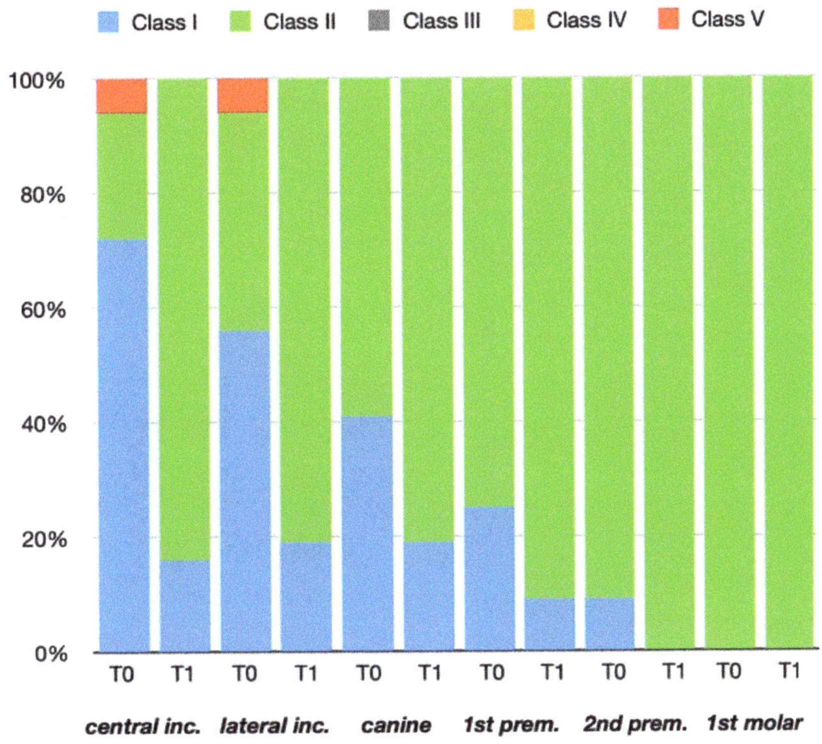

Figure 5. Percentage of sagittal root positions before and after therapy in the lower arch.

Regarding the lateral incisors and canines, 81% of the roots were centered in the alveolus at T1, although 19% remained in contact with the labial cortical plate.

The mandibular bicuspids and molars achieved a correct root position at the end of treatment: 91% of the first premolars and 100% of the roots of the second premolars and first molars were positioned between the two cortical plates. We can notice that the initial root position was better in the posterior than in the anterior teeth.

In the lower arch, no roots were found positioned outside the buccal cortical plate at T1.

By comparing Figures 4 and 5, the control of root position after clear aligner therapy was more effective in the upper arch, especially in the anterior sectors.

Subsequently, the torque values before and after therapy were evaluated and compared.

Table 2 illustrates the torque variation of the upper teeth due to the orthodontic treatment. At the end of therapy, the highest mean torque changes were achieved in the incisors, while the lowest mean torque changes were in the posterior teeth.

Table 2. Torque variations and paired t-test in the upper teeth ($\alpha = 0.05$).

Tooth	Torque Variation (°) (Mean ± SD)	Range (°)	Torque Variation (%) (Mean ± SD)	Range (%)	Tp	Confidence Interval	p-Value
Central incisor	3.26 ± 1.95	9.50–0.10	22.82 ± 12.62	66.66–1.20	12.49	3.78–2.74	*
Lateral incisor	2.57 ± 1.86	7.50–0.40	17.74 ± 9.09	48.68–7.03	10.31	3.07–2.07	*
Canine	1.53 ± 0.89	3.90–0	14.38 ± 6.52	32.05–0	12.84	1.77–1.29	*
First premolar	0.53 ± 0.36	1.10–0	8.59 ± 6.27	20.75–0	8.24	0.65–0.39	*
Second premolar	0.02 ± 0.07	0.20–−0.20	0.25 ± 0.91	2.74–−2.70	1.53	0.04–0	NS
First molar	0.01 ± 0.05	0.10–−0.20	0.15 ± 0.73	1.49–−2.70	1.28	0.03–0	NS

SD indicates standard deviation; and NS is not significant. * $p < 0.05$.

The torque increase was less noticeable in the lateral incisors and canines than in the central incisors. The mean percentage of torque increment varied from 0.15 ± 0.73% in the first molars to 22.82 ± 12.62% in the central incisors. This appears to agree with the variations of the root positions at T1 (Figure 4).

In the mandibular arch, the torque variations were the greatest in the central incisors, decreased in the lateral incisors and canines, and were minimal in the premolars and molars, as shown in Table 3.

Table 3. Torque variations and paired t-test in the lower teeth ($\alpha = 0.05$).

Tooth	Torque Variation (°) (Mean ± SD)	Range (°)	Torque Variation (%) (Mean ± SD)	Range (%)	Tp	Confidence Interval	p-Value
Central incisor	2.97 ± 2.53	9.00–0.20	22.63 ± 11.58	67.92–3.17	8.78	3.65–2.29	*
Lateral incisor	2.07 ± 2.16	8.10–0	12.95 ± 11.09	54.69–0	7.15	2.65–1.49	*
Canine	0.94 ± 0.95	3.40–0	7.82 ± 8.16	33.33–0	7.43	1.19–0.69	*
First premolar	0.18 ± 0.13	0.50–0	1.92 ± 1.67	6.25–0	7.67	0.26–0.13	*
Second premolar	0.02 ± 0.07	0.10–−0.20	0.18 ± 0.79	1.41–−2.47	1.31	0.04–0	NS
First molar	0.02 ± 0.06	0.10–−0.10	0.21 ± 0.84	1.30–−1.43	1.65	0.04–0	NS

SD indicates standard deviation; and NS is not significant. * $p < 0.05$.

The mean increase among mandibular teeth varied between 2.97 ± 2.53° (central incisors) and 0.02 ± 0.07°/0.02 ± 0.06° (second premolars/first molars).

Overall, the torque increase in the lower arch was less than in the upper arch, decreasing from the anterior to posterior sectors.

The results of the paired t-test are shown in Tables 2 and 3 for the maxillary and mandibular arches, respectively. In each arch, there was a statistically significant torque increase for the incisors, canines, and first premolars at the end of therapy.

Tables 4 and 5 summarize the differences between the maxillary and mandibular arches' initial and final root lengths.

Table 4. Root length changes and paired t-test in the upper arch ($\alpha = 0.05$).

Tooth	Changes in Root Length (mm) (Mean ± SD)	Range (mm)	Changes in Root Length (%) (Mean ± SD)	Range (%)	Tp	Confidence Interval	p-Value
Central incisor	−0.43 ± 0.27	−1.10–0.10	−3.27 ± 2.07	−8.53–0.76	8.92	0.53–0.33	*
Lateral incisor	−0.53 ± 0.38	−1.50–0.10	−4.03 ± 2.82	−10.79–0.78	7.88	0.66–0.39	*
Canine	−0.14 ± 0.13	−0.60–0	−0.83 ± 0.75	−3.57–0	6.28	0.19–0.09	*
First premolar	−0.08 ± 0.66	−0.30–0	−0.56 ± 0.42	−2.20–0	8.31	0.11–0.10	*
Second premolar	−0.003 ± 0.02	−0.10–0	−0.02 ± 0.13	−0.70–0	1	0.004–0	NS
First molar	−0.003 ± 0.01	−0.10–0	−0.02 ± 0.12	−0.68–0	1	0.004–0	NS

SD indicates standard deviation; and NS, is not significant. * $p < 0.05$.

Table 5. Root length changes and paired *t*-test in the lower arch (α = 0.05).

Tooth	Changes in Root Length (mm) (Mean ± SD)	Range (mm)	Changes in Root Length (%) (Mean ± SD)	Range (%)	Tp	Confidence Interval	*p*-Value
Central incisor	−0.20 ± 0.28	−0.80–0	−1.58 ± 1.77	−6.20–0	4.98	0.28–0.12	*
Lateral incisor	−0.09 ± 0.11	−0.50–0	−0.67 ± 0.79	−3.60–0	4.71	0.13–0.05	*
Canine	−0.08 ± 0.07	−0.20–0	−0.50 ± 0.41	−1.31–0	6.70	0.10–0.05	*
First premolar	−0.003 ± 0.01	−0.10–0	−0.02 ± 0.13	−0.71–0	1	0.004–0	NS
Second premolar	−0.003 ± 0.02	−0.10–0	−0.02 ± 0.13	−0.75–0	1	0.004–0	NS
First molar	−0.003 ± 0.02	−0.10–0	−0.03 ± 0.14	−0.80–0	1	0.004–0	NS

SD indicates standard deviation; and NS is not significant. * $p < 0.05$.

In the upper arch, the most significant mean root reduction was detected in the lateral incisors (−0.53 ± 0.38 mm), followed by the central incisors (−0.43 ± 0.27 mm) and canines (−0.14 ± 0.13 mm). The root length changes were minimal on the first premolars, while they were negligible for the second premolars and first molars.

According to the severity of EARR, all upper teeth showed mild resorption (<10%), except for two upper lateral incisors, which have undergone moderate resorption (10.79% and 10.23%).

As regards the lower arch, the root length loss was more significant in the central incisors, and then decreased in the lateral incisors and canines and became negligible for the premolars and first molars.

The mean absolute reduction in the mandibular arch varied from 0.003 ± 0.02 mm (second premolar and first molar) to 0.20 ± 0.28 mm (central incisor).

Neither moderate nor severe resorption was found in any lower tooth, which showed a resorption below 6.20%.

The root length decrease was statistically significant on the maxillary incisors, canines, first premolars (Table 4), and mandibular incisors and canines (Table 5).

4. Disscussion

This research aimed to study the efficacy of clear aligners regarding root movement by evaluating the torque before and after treatment and verifying whether the torque variations led to root resorption.

One of the significant challenges for clear aligners is controlling root movement, including the labiolingual inclination, which is more effective in the upper arch than in the lower one. This may suggest that the inter-proximal enamel reduction is a reasonable solution to resolve lower crowding and preserve the roots' integrity.

The torque variation was, as previously reported, higher and statistically significant in the upper and lower incisors: in these teeth, the mean torque variation was more important than 2°.

The torque increase in the canines was lower than in the incisors: it can be explained by their position at the curvature of the dental arch, which does not allow them to receive optimal force. Moreover, the mechanical efficiency for delivering effective buccally directed force by the aligner decreases from anterior to posterior sectors [23], which is mainly related to more complex root anatomy, thicker cortical plate, higher mastication loading, and greater soft tissue resistance from the cheeks in the posterior region [24].

In literature, the outcomes regarding the torque variation appear to be discordant: this is due to methodological heterogeneity. Some studies analyzed torque without planning for optimized attachments, auxiliaries, or power ridges [25]; others used different materials, even materials no longer used today [1]. Different material properties and aligner production processes affect the force levels and, thus, the predictability of tooth movements [26].

Although clear aligners seem to meet all the criteria of an ideal orthodontic appliance, some biomechanical limits have yet to be overcome [27]. The predictability of tooth movement widely depends on different types of the tooth, dental movement, and arch.

The aligners show the same biomechanical principles as other orthodontic appliances, but the material properties of clear aligners are probably responsible for their weak accuracy in applying torque. Indeed, an aligner's gingival margin is elastic, which would have difficulty controlling forces applied in this region [28]. To overcome this limit, some manufacturers have introduced power ridges, which generate a more significant moment than attachments, as described by Castroflorio et al. [29]: this represents a valid alternative for accurate root torque control, especially on the incisors.

Therefore, the aligners with power ridges may determine a better expression of the torque than can be achieved compared to the multibracket appliances, at least in some prescriptions; moreover, in fixed therapy, Morina et al. [30] showed that various 0.022″ self-ligating and conventional brackets on the upper incisors lost an average 10° of torque, known as the so-called "torque play," after the insertion of 0.019″ × 0.025″ arch wires.

Previous studies [31,32] proved that a clear aligner could produce clinically acceptable results compared to conventional and self-ligating brackets. No statistically significant differences were seen among the three mechanotherapy.

The possibility of managing root movement may help us to correctly plan for the position of the roots within the cortical plates. Only today, the virtual setups have displayed only the coronal changes rather than the radicular ones, which does not accurately reflect the patient's final occlusion.

In addition, there are studies [1] which evaluated the torque only on a small number of aligners and, for this reason, concluded that a greater number of aligners or a torque overcorrection should be required to achieve the desired dental movements. In the present research, the refinements and power ridges have also been included.

It is essential to underline those previous studies concerning torque changes and root resorptions that limited their interest only to the anterior teeth of the upper arch. In the current study, the torque variation and root length loss have been evaluated for both the upper and lower teeth, from incisors to molars.

The subjects selected for this study were adults between 18 and 38 years old. Therefore, the influence of bone metabolism on the movement of teeth during puberty and perimenopause was eliminated.

In our study, comprehensive treatment with clear aligners has led to mild root resorptions; indeed, only two upper lateral incisors showed moderate resorption.

The difference in root length between before and after therapy was found to be statistically significant for all upper and lower anterior teeth; in the posterior sectors, on the other hand, there was no significant root length loss, and this appears to agree with the small variations of root position in the second premolars and molars between before and after therapy.

Data from the literature agree with those found in the current study, which extended its evaluation to the lower arch and the posterior teeth. Liu et al. [33] noticed that most incisors of their sample showed mild or moderate resorption, and only a small percentage (0.625%) underwent severe resorption.

The current study investigated only patients who received a non-extraction treatment; the incidence of EARR was lower in non-extraction than in extraction cases due to the large dental movement and reduction in overjet when closing extraction spaces [34].

Crowding is another risk factor for EARR, with the percentage of root length change in mild crowding significantly lower than in severe crowding.

Root resorption was more related to intrusive and extrusive forces; vertical movement, especially of the anterior teeth, produced greater stress at the root apex [35].

Consequently, the orthodontist should pay attention to the initial torque and tooth movement during the elaboration of the treatment plan.

Regarding the treatment duration, whether this could be a risk factor remains controversial.

In the current study, although the average duration of treatment was 25 months, we found mild root resorption. This can be explained by the intermittent and light force in clear aligner therapy, and by pauses in treatment. At the same time, refinements to a patient's unique prescription are being made, providing the potential for cementum healing [36]. Consequently, the period between therapy initiation and termination does not represent the active treatment duration.

According to the literature, the prevalence of severe resorption with multibracket appliances is higher than in clear aligners [9–37]; fortunately, the long-term survival of teeth with severe root resorption appears to be good.

Regarding the instrumental examination, we used CBCT scan with the low dose protocol because panoramic and periapical radiographs tend to underestimate the finding of root resorption by more than 20% compared to CBCT [38]; indeed, CBCTs show high sensitivity, specificity, and reproducibility.

In orthodontics, the 2D examinations, e.g. panoramic or lateral cephalometric radiographs, lack all information required during diagnosis, treatment planning, and follow-up. In contrast, CBCT scans provide detailed 3D images on skeletal tissue, root resorptions, facial muscle measurement (e.g. masseter), TMJ morphology, and upper airway status.

To limit patient radiation exposure, low-dose CBCT protocols have been proposed. In the current study, we used a low-dose CBCT with a value of the effective dose of 35 μSv [17], considerably lower than that of traditional CBCT.

There is a need for clinical recommendations, guidelines, or position statements from authoritative bodies regarding the use of the low-dose protocol in dentistry. The clinical recommendations issued by the American Academy of Oral and Maxillofacial Radiology (AAOMR) [39] in 2013 concern only high-dose CBCT. In addition, according to AAOMR, in orthodontics, the recommendations are "neither rigid guidelines nor did they represent or imply a standard of care"; therefore, the orthodontist evaluates the use of CBCT based on clinical presentation considering the patient's best interest. Recently, Yeung et al. [40] in their review have underlined that the low-dose CBCT undoubtedly offers greater information both at the beginning and end of the therapy in various dental medicine disciplines, such as in orthodontics to evaluate impacted teeth or alveolar bone quantity and to assess TMJs, in endodontics to detect root fractures, resorptions, and periapical bone loss, and in dental surgery to plan implant insertion and to place temporary anchorage devices.

The current study was subjected to the following limitations. Firstly, we examined only non-extraction treatment. In extraction cases, the management of torque during closing extraction spaces is challenging for the orthodontist [41], and consequently, the root length loss is higher. Secondly, we excluded the use of intermaxillary elastics, so the incidence of EARR was underestimated in our study since the intermittent forces of intermaxillary elastics could lead to greater root resorption. Lastly, we considered only aligners produced by a single manufacturer, and aligner materials' clinical performances vary from manufacturer to manufacturer.

Future papers are suggested to establish whether the materials of other clear aligner manufacturers show similar outcomes to those reported in the present study.

We found, within study limitations, an improvement of sagittal root position and a significant increase in torque in the upper and lower incisors, canines, and first premolars, as well as the highest mean torque changes in anterior teeth. Moreover, these torque increases led to no severe root resorptions, and we observed mainly mild root resorptions following clear aligner treatment.

Therefore, compared to previous papers, our study extended the assessment of root control management to all upper and lower teeth and demonstrated that the clear aligners could enhance the torque in both arches without a substantial root length loss.

5. Conclusions

The introduction of clear aligner therapy offers an additional therapeutic tool for the resolution of mild to moderate malocclusions, available for patients, especially adults, who are attentive to the aesthetics of their smile or who, for personal needs, require little visible and non-altering phonation appliances. Furthermore, the removable aligners allow the maintenance of a correct home and professional oral hygiene.

In the present study, aligners showed an improvement in sagittal root position and torque, especially in the upper anterior region.

To manage the correct root position during clear aligner therapy, the orthodontist should evaluate some factors, such as the staging (movement per aligner), the frequency of aligner changing, the severity of malocclusion, the clinician's experience, the morphology and position of the attachments, and the use of auxiliaries, which could affect the success of the treatment.

Regarding root resorption, most teeth showed mild resorption, and our findings agree with the previous studies. It would be interesting to assess whether the type of attachments used, and the specific tooth movements influence the amount of root resorption.

Finally, in daily practice, it would be desirable to utilize software that, by integrating the three-dimensional data from CBCT, also allows a radicular virtual setup to control root position and prevent apical root resorption and periodontal disease.

Author Contributions: Methodology, M.M. and F.F.; Software, S.M.; Formal analysis, M.M.; Investigation, M.M. and G.V.; Data curation, M.M.; Writing—original draft, M.M.; Writing—review & editing, M.M. All authors have read and agreed to the published version of the manuscript.

Funding: This research received no external funding.

Institutional Review Board Statement: Ethics approval (number 23) was obtained by the hospital's Independent Ethics Committee of Chieti. The study protocol was drawn following the European Union Good Practice Rules, and the Helsinki Declaration.

Data Availability Statement: The data presented in this study are available on request from the corresponding author. The data are not publicly available due to privacy.

Conflicts of Interest: The authors declare no conflict of interest.

References

1. Simon, M.; Keilig, L.; Schwarze, J.; Jun, B.A.; Bourauel, C. Treatment outcome and efficacy of an aligner technique–regarding incisor torque, premolar derotation and molar distalization. *BMC Oral Health* **2014**, *11*, 14–68. [CrossRef] [PubMed]
2. Jiang, T.; Jiang, Y.; Chu, F.; Lu, P.; Tang, G. A cone-beam computed tomographic study evaluating the efficacy of incisor movement with clear aligners: Assessment of incisor pure tipping, controlled tipping, translation, and torque. *Am. J. Orthod. Dentofac. Orthop.* **2021**, *159*, 635–643. [CrossRef] [PubMed]
3. Lacruz, R.S.; Bermúdez de Castro, J.M.; Martinón-Torres, M.; O'Higgins, P.; Paine, M.L.; Carbonell, E.; Arsuaga, J.L.; Bromage, T.G. Facial morphogenesis of the earliest europeans. *PLoS ONE* **2013**, *6*, e65199. [CrossRef] [PubMed]
4. Macrì, M.; Festa, F. Three-dimensional evaluation using CBCT of the mandibular asymmetry and the compensation mechanism in a growing patient: A case report. *Front. Public Health* **2022**, *10*, 921413. [CrossRef] [PubMed]
5. Archambault, A.; Major, T.W.; Carey, J.P.; Heo, G.; Badawi, H.; Major, P.W. A comparison of torque expression between stainless steel, titanium molybdenum alloy and copper nickel titanium wires in metallic self-ligating brackets. *Angle Orthod.* **2010**, *80*, 884–889. [CrossRef]
6. Rauch, D.E. Torque and its application to orthodontics. *Am. J. Orthod.* **1959**, *45*, 817–830. [CrossRef]
7. Hemingway, R.; Williams, R.L.; Hunt, J.A.; Rudge, S.J. The influence of bracket type on the force delivery of Ni-Ti archwires. *Eur. J. Orthod.* **2001**, *23*, 233–241. [CrossRef]
8. Elhaddaoui, R.; Qoraich, H.S.; Bahije, L.; Zaoui, F. Orthodontic aligners and root resorption: A systematic review. *Int. Orthod.* **2017**, *15*, 1–12. [CrossRef]
9. Li, Y.; Deng, S.; Mei, L.; Li, Z.; Zhang, X.; Yang, C.; Li, Y. Prevalence and severity of apical root resorption during orthodontic treatment with clear aligners and fixed appliances: A cone beam computed tomography study. *Prog. Orthod.* **2020**, *21*, 1. [CrossRef]
10. Krieger, E.; Drechsler, T.; Schmidtmann, I.; Jacobs, C.; Haag, S.; Wehrbein, H. Apical root resorption during orthodontic treatment with aligners? A retrospective radiometric study. *Head Face Med.* **2013**, *9*, 21. [CrossRef]
11. Iglesias-Linares, A.; Sonnenberg, B.; Solano, B.; Yañez-Vico, R.M.; Solano, E.; Lindauer, S.J.; Flores-Mir, C. Orthodontically induced external apical root resorption in patients treated with fixed appliances vs removable aligners. *Angle Orthod.* **2017**, *87*, 3–10. [CrossRef] [PubMed]

12. Lascala, C.A.; Panella, J.; Marques, M.M. Analysis of the accuracy of linear measurements obtained by cone beam computed tomography (CBCT-NewTom). *Dentomaxillofac. Radiol.* **2004**, *33*, 291–294. [CrossRef]
13. Dudic, A.; Giannopoulou, C.; Martinez, M.; Montet, X.; Kiliaridis, S. Diagnostic accuracy of digitized periapical radiographs validated against micro-computed tomography scanning in evaluating orthodontically induced apical root resorption. *Eur. J. Oral Sci.* **2008**, *116*, 467–472. [CrossRef]
14. Lund, H.; Gröndahl, K.; Gröndahl, H. Cone beam computed tomography for assessment of root length and marginal bone level during orthodontic treatment. *Angle Orthod.* **2010**, *80*, 466–473. [CrossRef]
15. Ren, H.; Chen, J.; Deng, F.; Zheng, L.; Liu, X.; Dong, Y. Comparison of cone-beam computed tomography and periapical radiography for detecting simulated apical root resorption. *Angle Orthod.* **2013**, *83*, 189–195. [CrossRef]
16. Festa, F.; Rotelli, C.; Scarano, A.; Navarra, R.; Caulo, M.; Macrì, M. Functional Magnetic Resonance Connectivity in Patients with Temporomadibular Joint Disorders. *Front. Neurol.* **2021**, *12*, 629211. [CrossRef]
17. Feragalli, B.; Rampado, O.; Abate, C.; Macrì, M.; Festa, F.; Stromei, F.; Caputi, S.; Guglielmi, G. Cone beam computed tomography for dental and maxillofacial imaging: Technique improvement and low-dose protocols. *Radiol. Med.* **2017**, *122*, 581–588. [CrossRef] [PubMed]
18. Nucera, R.; Ciancio, E.; Maino, G.; Barbera, S.; Imbesi, E.; Bellocchio, A.M. Evaluation of bone depth, cortical bone, and mucosa thickness of palatal posterior supra-alveolar insertion site for miniscrew placement. *Prog. Orthod.* **2022**, *23*, 18. [CrossRef] [PubMed]
19. Macrì, M.; Toniato, E.; Murmura, G.; Varvara, G.; Festa, F. Midpalatal Suture Density as a Function of Sex and Growth-Pattern-Related Variability via CBCT Evaluations of 392 Adolescents Treated with a Rapid Maxillary Expander Appliance. *Appl. Sci.* **2022**, *12*, 2221. [CrossRef]
20. Kan, J.Y.; Roe, P.; Rungcharassaeng, K.; Patel, R.D.; Waki, T.; Lozada, J.L. Classification of sagittal root position in relation to the anterior maxillary osseous housing for immediate implant placement: A cone beam computed tomography study. *Int. J. Oral Maxillofac. Implant.* **2011**, *26*, 873–876.
21. Aman, C.; Azevedo, B.; Bednar, E.; Chandiramami, S.; German, D.; Nicholson, E.; Nicholson, K.; Scarfe, W.C. Apical root resorption during orthodontic treatment with clear aligners: A retrospective study using cone-beam computed tomography. *Am. J. Orthod. Dentofac. Orthop.* **2018**, *153*, 842–851. [CrossRef] [PubMed]
22. Tong, H.; Enciso, R.; Van Elslande, D.; Major, P.W.; Sameshima, G.T. A new method to measure mesiodistal angulation and faciolingual inclination of each whole tooth with volumetric cone-beam computed tomography images. *Am. J. Orthod. Dentofac. Orthop.* **2012**, *142*, 133–143. [CrossRef] [PubMed]
23. Zhou, N.; Guo, J. Efficiency of upper arch expansion with the Invisalign system. *Angle Orthod.* **2020**, *90*, 23–30. [CrossRef] [PubMed]
24. Houle, J.P.; Piedade, L.; Todescan, R., Jr.; Pinheiro, F.H. The predictability of transverse changes with Invisalign. *Angle Orthod.* **2017**, *87*, 19–24. [CrossRef] [PubMed]
25. Zhang, X.J.; He, L.; Guo, H.M.; Tian, J.; Bai, Y.X.; Li, S. Integrated three-dimensional digital assessment of accuracy of anterior tooth movement using clear aligners. *Korean J. Orthod.* **2015**, *45*, 275–281. [CrossRef] [PubMed]
26. Macrì, M.; Murmura, G.; Varvara, G.; Traini, T.; Festa, F. Clinical Performances and Biological Features of Clear Aligners Materials in Orthodontics. *Front. Mater.* **2022**, *9*, 819121. [CrossRef]
27. Schupp, W.; Haubrich, J.; Neumann, I. Class II correction with the Invisalign system. *J. Clin. Orthod.* **2010**, *44*, 28–35. [PubMed]
28. Hahn, W.; Zapf, A.; Dathe, H.; Fialka-Fricke, J.; Fricke-Zech, S.; Gruber, R.; Kubein-Meesenburg, D.; Sadat-Khonsari, R. Torquing an upper central incisor with aligners-acting forces and biomechanical principles. *Eur. J. Orthod.* **2010**, *32*, 607–613. [CrossRef]
29. Castroflorio, T.; Garino, F.; Lazzaro, A.; Debernardi, C. Upper-incisor root control with Invisalign appliances. *J. Clin. Orthod.* **2013**, *47*, 346–351.
30. Morina, E.; Eliades, T.; Pandis, N.; Jäger, A.; Bourauel, C. Torque expression of self-ligating brackets compared with conventional metallic, ceramic, and plastic brackets. *Eur. J. Orthod.* **2008**, *30*, 233–238. [CrossRef]
31. Hennessy, J.; Garvey, T.; Al-Awadhi, E.A. A randomized clinical trial comparing mandibular incisor proclination produced by fixed labial appliances and clear aligners. *Angle Orthod.* **2016**, *86*, 706–712. [CrossRef] [PubMed]
32. Sfondrini, M.F.; Gandini, P.; Castroflorio, T.; Garino, F.; Mergati, L.; D'Anca, K.; Trovati, F.; Scribante, A. Buccolingual inclination control of upper central incisors of aligners: A comparison with conventional and self-ligating brackets. *Biomed. Res. Int.* **2018**, *2018*, 9341821. [CrossRef] [PubMed]
33. Liu, W.; Shao, J.; Li, S.; Al-Balaa, M.; Xia, L.; Li, H.; Hua, X. Volumetric cone-beam computed tomography evaluation and risk factor analysis of external apical root resorption with clear aligner therapy. *Angle Orthod.* **2021**, *91*, 597–603. [CrossRef] [PubMed]
34. Pastro, J.D.V.; Nogueira, A.C.A.; Salvatore de Freitas, K.M.; Valarelli, F.P.; Cançado, R.H.; de Oliveira, R.C.G.; de Oliveira, R.C.G. Factors associated to apical root resorption after orthodontic treatment. *Open Dent. J.* **2018**, *12*, 331–339. [CrossRef] [PubMed]
35. Rudolph, D.J.; Willes, M.G.; Sameshima, G.T. A finite element model of apical force distribution from orthodontic tooth movement. *Angle Orthod.* **2001**, *71*, 127–131. [CrossRef]
36. Lund, H.; Gröndahl, K.; Hansen, K.; Gröndahl, H.G. Apical root resorption during orthodontic treatment. A prospective study using cone beam CT. *Angle Orthod.* **2012**, *82*, 480–487. [CrossRef]
37. Eissa, O.; Carlyle, T.; El-Bialy, T. Evaluation of root length following treatment with clear aligners and two different fixed orthodontic appliances. A pilot study. *J. Orthod. Sci.* **2018**, *7*, 11. [CrossRef]

38. Dudic, A.; Giannopoulou, C.; Leuzinger, M.; Kiliaridis, S. Detection of apical root resorption after orthodontic treatment by using panoramic radiography and cone-beam computed tomography of super-high resolution. *Am. J. Orthod. Dentofac. Orthop.* **2009**, *135*, 434–437. [CrossRef]
39. American Academy of Oral and Maxillofacial Radiology. Clinical recommendations regarding use of cone beam computed tomography in orthodontics. [corrected]. Position statement by the American Academy of Oral and Maxillofacial Radiology. *Oral Surg. Oral Med. Oral Pathol. Oral Radiol.* **2013**, *116*, 238–257. [CrossRef]
40. Yeung, A.W.; Jacobs, R.; Bornstein, M.M. Novel low-dose protocols using cone beam computed tomography in dental medicine: A review focusing on indications, limitations, and future possibilities. *Clin. Oral Investig.* **2019**, *23*, 2573–2581. [CrossRef]
41. Cheng, Y.; Gao, J.; Fang, S.; Wang, W.; Ma, Y.; Jin, Z. Torque movement of the upper anterior teeth using a clear aligner in cases of extraction: A finite element study. *Prog. Orthod.* **2022**, *23*, 26. [CrossRef] [PubMed]

applied sciences

MDPI

Article

Virtual Surgical Reduction in Atrophic Edentulous Mandible Fractures: A Novel Approach Based on "in House" Digital Work-Flow

Vincenzo Abbate [1,*], Umberto Committeri [1], Stefania Troise [1], Paola Bonavolontà [1], Luigi Angelo Vaira [2], Guido Gabriele [3], Federico Biglioli [4], Filippo Tarabbia [4], Luigi Califano [1] and Giovanni Dell'Aversana Orabona [1]

1 Maxillofacial Surgery Unit, Department of Neurosciences, Reproductive and Odontostomatological Sciences, University of Naples Federico II, Via Sergio Pansini 5, 80131 Naples, Italy
2 Maxillofacial Surgery Operative Unit, University Hospital of Sassari, Viale San Pietro 43/B, 07100 Sassari, Italy
3 Maxillofacial Surgery Operative Unit, University Hospital of Siena, Via Banchi di Sotto 55, 53100 Siena, Italy
4 Maxillofacial Surgery Operative Unit, ASST Santi Paolo e Carlo, Via Antonio di Rudinì, 8, 20142 Milano, Italy
* Correspondence: vincenzo.abbate@unina.it; Tel.: +39-3357-266-177

Featured Application: Virtual Surgical Planning in Cranio-Facial Traumatology.

Abstract: Atrophic edentulous mandible fractures are a challenge for maxillo-facial surgeons because of low vascularization, low bone regeneration, and lack of occlusion. Whereas occlusion is the main guide in the reduction of mandibular fractures, the aim of our study is to show the advantages of using virtual surgical planning (VSP) in surgery when the occlusal guide is absent. This work is a prospective study that shows the in-house digital workflow for the management of these fractures in the Maxillo-Facial Surgery Unit of Federico II University Hospital of Naples. Four patients who satisfied the criteria were included in the study. For each patient, the same defined CAD/CAM-based was applied. The workflow followed four steps: (1) bone segmentation and virtual reduction of fracture fragments; (2) three-dimensional printing of virtually reduced mandible and modelling of 2.4 reconstruction plate on printed resin model; (3) surgery aided by the pre-formed plate; (4) digital and clinical outcomes analysis. In the last step, a distance colour map was conducted to compare the virtual planning and postoperative CT outcome. In all cases, the discrepancies values between the two images were lower than 1.5 mm, and good clinical outcomes in terms of facial symmetry, absence of sensory disturbance, and possibility of prosthetic rehabilitation were obtained. In conclusion, the VSP, with our in-house workflow brings benefits in the management of atrophic edentulous mandible fractures in terms of the high accuracy of bone repositioning.

Keywords: atrophic edentulous mandible; mandibular fractures; virtual surgical planning; reconstruction-plate; digital workflow; CAD/CAM technology

Citation: Abbate, V.; Committeri, U.; Troise, S.; Bonavolontà, P.; Vaira, L.A.; Gabriele, G.; Biglioli, F.; Tarabbia, F.; Califano, L.; Dell'Aversana Orabona, G. Virtual Surgical Reduction in Atrophic Edentulous Mandible Fractures: A Novel Approach Based on "in House" Digital Work-Flow. *Appl. Sci.* **2023**, *13*, 1474. https://doi.org/10.3390/app13031474

Academic Editor: Bruno Chrcanovic

Received: 7 December 2022
Revised: 17 January 2023
Accepted: 19 January 2023
Published: 22 January 2023

1. Introduction

The incidence of fractures in atrophic edentulous mandibles has been reported to range between 1% to 5% of all mandibular traumas; in elderly people, these fractures range between 10.1% to 56% due to poor proprioception, weakness, and impaired reflexes, which lead to a greater frequency of accidental falls. The indications to treat these fractures are mainly based on the possibility to guarantee a good quality of life in terms of chewing, swallowing, phonation, displacing of bone stumps, and the possibility of prosthetic rehabilitation [1]. Therefore, the goal of surgical treatment is to restore anatomical continuity and facial symmetry to obtain valuable prosthetic rehabilitation. Open Reduction Internal Fixation (ORIF) represents the most valuable approach [2–5], but the management of such kinds of fractures is still a challenge for the technical difficulties and the comorbidities that may afflict elderly patients.

In such cases, the fragments' anatomical reduction and fracture consolidation are difficult to achieve due to bone atrophy which lowers bone regeneration and causes poor vascularization regularly guaranteed by the periosteum. Thus, a load-bearing plate is mandatory due to the decrease in bony volume and vascularization in these high-risk patients. However, the lack of occlusions is the main problem in these fractures since occlusions are the main guide for a stable reduction. Surgeons face difficulties in operating with no references to align the bone stumps [6,7] in patients who lack dental elements. To solve these difficulties, computer-assisted design (CAD) and computer-assisted manufacturing (CAM) technology are increasingly used in clinical practice. The ability to virtually program surgical procedures and transfer the planning to the operating room is now finding more and more applications in maxillofacial surgery [8,9]. The main application fields of VSP in maxillofacial surgery were orthognathic/malformation surgery, implant surgery, and oncological-reconstructive surgery. The introduction of VSP into craniofacial traumatology is certainly more recent and it has been mainly documented for complex fractures, especially for comminuted fractures, and in cases with fragments displacement. The efficacy of the analogical procedure has been confirmed for orbital, zygomatic, and dentate mandible fractures, but there is not sufficient evidence regarding fractures of edentulous atrophic mandibles. Based on the previous evidence, CAD/CAM (or VSP) technologies could be a support to the surgeon in the treatment of these complex fractures. Hence, the scope of this paper is to present a new in-house digital workflow aimed to finalize a patient-specific implant (PSI) as a reference for effective Open Reduction and Internal Fixation (ORIF) when the occlusal guide is absent.

2. Materials and Methods

This work is a prospective study that shows the in-house digital workflow for the management of complex fractures in atrophic edentulous mandibles. The study was conducted in the period between October 2020 and March 2021 on patients admitted to the Maxillo-Facial Surgery Unit of Federico II University Hospital in Naples. The study met the criteria established by the Declaration of Helsinki and was approved by the Ethics Committee of the Federico II University hospital of Naples with protocol number 373/19. The patients enrolled in the study satisfied the following inclusion criteria:

- post-traumatic comminuted mandibular fractures verified with a CT scan within 12 h from the trauma
- atrophic edentulous mandible
- unavailability of the patient's personal dental prostheses that could guide the occlusion; written informed consent to undergo the surgical procedure

Patients who met the following criteria were excluded:

- previous surgically treated mandibular fractures
- patients who have refused surgical treatment
- partial edentulism
- taking drugs that affect bone resorption

Among nine patients likely to be enrolled, only four patients satisfied the inclusion criteria and were enrolled in the study; two patients presented partial edentulism but enough to allow a fracture reduction based on the occlusion; one patient had tumour pathologies and took bisphosphonates medications, causing a poor bone quality; one patient refused surgical treatment; one patient previously underwent surgical treatment for mandibular neoplasia.

Among the four patients enrolled in the study, a computer-assisted digital workflow was settled to perform the virtual reduction of the fractures and to finalize a patient specific plate (PSP) to fix the mandible. For each patient, the same CAD/CAM based workflow was applied as defined below:

(1) Digital Imaging and Communications in Medicine (DICOM) to Stereolythography (STL) files: by using Materialize Mimics Medical 21.0 software (Technologielaan 15,

3001 Leuven, Belgium) to perform the virtual reduction. The first step was to upload the DICOM-Files to generate a 3D model of the patient. Then, the segmentation of bone fragments was possible thanks to the New Mask tool. Generally, the range of bone threshold was between 0–1250 HU. Once the first step was completed, a three-dimensional image of the entire splanchnocranium was obtained. The division of each bone fragment was needed to proceed with the simulation of the fracture reduction. Therefore, in the second step, the Split Mask tool allowed to divide of the main mask into single segments for each bony fragment. In this way, each fragment could be repositioned virtually mimicking the fracture reduction, while the optional function Smart Fill was used to fill the bone cavities. The process of identification of the fracture lines was performed by three different maxillo-facial surgeons (VA, UC and ST). The produced images were then converted into "objects files". Using the 3-Matic Medical 3D (Materialise, Leuven, BE) it was possible to convert the stumps into a 3D plan and obtain a virtual reduction with the stumps in the correct anatomical position. To avoid any bias, three different maxillofacial surgeons (VA, UC and ST) also performed the procedure. The reduction was obtained by considering the alignment of the condyles in the glenoid fossa. At the end of the procedure, the virtually reduced mandible file was converted into a STL file adapted for the "Formlab-Form 3B" 3D printer (Figure 1a,b).

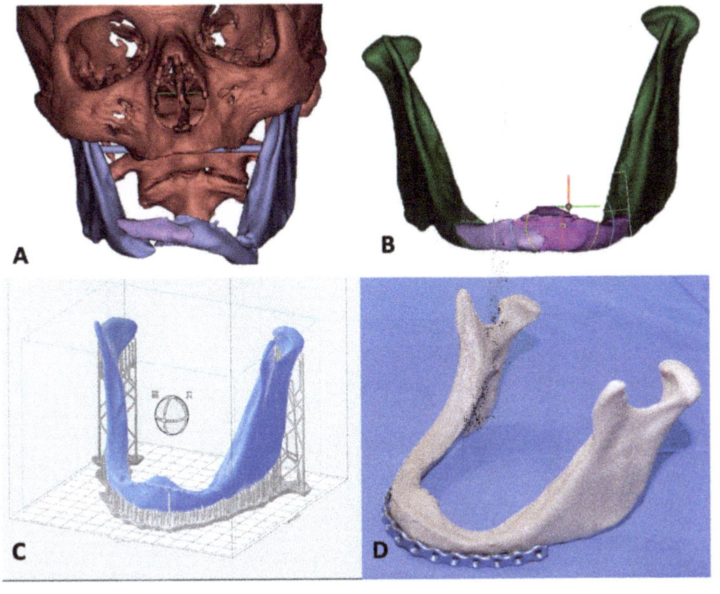

Figure 1. "In-house" workflow. (**A**) Segmentation of the fracture segments in the pre-operative CT using the Materialize Mimics 21.0 Medical Software; (**B**) virtual reduction of fracture fragments; (**C**) 3D printing setting of the virtual planned reduced mandible using the Preform 3D Printing Software; (**D**) Modeling of the plate on the reduced mandible resin model.

(2) In House Rapid prototyping 3D Print—Resin model and 2.4 reconstruction plate modelling: a specific slicing software (PreForm 3D Printing Software—Formlabs 3B+) was used to set the model for 3D printing. It was necessary to use the supports to have the highest possible quality of the model (Figure 1c). A resin StereoLithography Apparatus (SLA) 3D printer (Formlab-Form 3B+, located in Somerville, MA, USA) was used for in-house rapid prototyping. The selected material was the Formlab Model Resin V3 (ISO 10993-5: 2009) because of its mechanical characteristics such as tensile strength of 27 MPa and modulus of elasticity of 1.1 GPa. Once the print was completed, the model was subjected to an autonomous immersion bath in isopropyl alcohol using the Form Wash

(Form Wash -located in Somerville, MA, USA) device for 20 min. The final step was the photopolymerization using the Form Cure (Form Cure located in Somerville, MA, USA) device that uses the UV order to increase the tensile strength up to 48 MPa and modulus of elasticity up to 2.3 GPa.

Once the reduced mandible model was obtained, a 2.4 reconstruction plate was modelled to be applied during the surgery (Figure 1d). The plate was bent by applying manual pressure to finalize a customized device for each patient and was sterilized in an autoclave with a temperature of 160 °C for 60 min, 24 h before surgery.

(3) Performing surgery: all the surgeries were performed by the same surgical team and under general anesthesia. A small cutaneous submental surgical access was performed for an extension of the fracture of about 6 cm in each operation. After soft tissue detachment, the bone fragments were identified, and the cortical bone was skeletonized to avoid the complete elevation of the periosteum and not to create further loss of the vascularization. Then, the plate was positioned between the two fixed bone stumps of mandibular body. The fragments were replaced in the correct anatomic position, and the reduction of the comminuted fracture of the mandible was possible based on the previously modeled plate. At the end of the procedure, the surgeon checked the facial symmetry, joint functions, and condyles correctly in the glenoid fossa bilaterally. A nylon 4.0 was used for the final suture. A surgical procedure example is shown in Figure 2a–d.

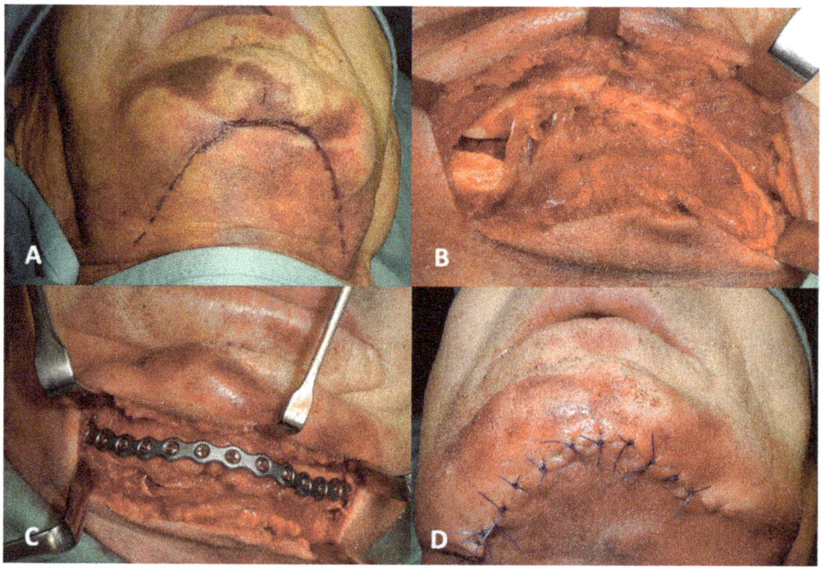

Figure 2. Surgery. (**A**): submental approach; (**B**): exposure of the fracture fragments; (**C**): ORIF of fracture with preformed plate; (**D**): suture in nylon.

(4) Initial digital outcomes: all the patients underwent a post-surgical CT scan the following day of the surgery. By using Materialize Mimics Medical 21.0 software, DICOM data have been converted into STL files. The workflow to obtain segmented mandible was the same described previously. The Geomagic Design X software (3D point cloud and mesh processing software, 11 Breedewues, L-1259 Senningerberg—Production, Logistics & Service Center) was used to compare the planned STL file in a 3D space obtained by our digital planning procedure with the post-surgical STL file. A distanced color map was obtained overlapping the two images and the discrepancies in mm between these two images in all four cases (Figure 3). To reduce any human errors in the overlapping process, the non-manual but automated protocol of the Geomagic software was used. The upper/lower limit for color coding of the discrepancies was fixed as +2 mm and

−2 mm, so that deviations appeared in different colors. Steps of 0.25 mm were used, so each color encoded a distance interval of 0.25 mm. A total discrepancy at the cloud point was calculated between the two images. However, to avoid the bias of the plate in the postoperative image the discrepancies were also calculated manually at the level of well defined bone points. Therefore, 11 anatomical bone landmarks were set to compare the "planned model" and the "in vivo" result on the following cephalometric points: menton, pogonion, B point, left and right condyle, left and right gonion, left and right lingula, left and right mental foramen.

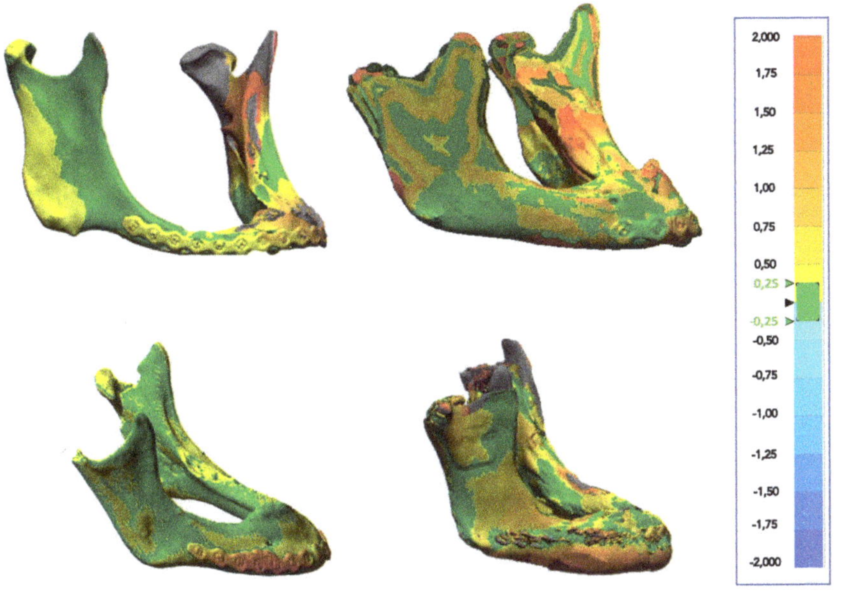

Figure 3. Digital analysis. Distance color map of the four cases. On the right the color legend bar.

(5) Long-term outcomes: all the patients were clinically evaluated during the outpatient follow-up. The patients went through periodic controls at 1 month and at six months period. During the outpatient controls aimed to check/control the facial symmetry, pain, joint and masticatory function, mouth opening, sensory, and motor alterations related to deficits of the V and VII cranial nerve.

3. Results

Four patients were treated using the CAD/CAM and 3D printing technology as the described protocol. The sex was equally distributed, the average age was 79 years (range 61–88 years). The operating times were very short, with an average surgical time of 45 min. No complications were observed during surgery or at the immediate post-surgical care. The average hospitalization time after surgery was four days (ranging from two to six days). At the six-monthly follow-up, we observed good clinical outcomes in our samples in terms of facial symmetry restoration, absence of sensory disturbances of the inferior alveolar nerve, no facial nerve weakness, satisfying joint and masticatory function, and mean mouth opening of 37 mm (rangin from 33 to 40 mm).

Regarding the digital workflow, the virtual planning design took around 2 h for each case while about 6 h for each case for the printing process and sterilization. On average, about 45 ± 2 mL of resin was required to print the models. An estimated cost of about 6.3 Euros is estimated for each case. The discrepancy values deriving from the overlapping analysis of the models were shown in Table 1 (Table 1). In the examined samples, all the obtained discrepancies' values were less than 1.5 mm. The mean value of the discrepancies of the whole overlapping cloud point was 0.69 ± 0.33 mm. Regarding the discrepancies of the cephalometric examined points in the sample, the higher mean value was obtained in the condylar site (Right: 0.80 ± 0.26 mm—Left: 0.79 ± 0.32 mm), while the lower mean value was obtained in the B Point (0.55 ± 0.41 mm).

Table 1. Discrepancies analysis.

	CASE 1	CASE 2	CASE 3	CASE 4	Average ± SD
	Discrepancy mm	**Discrepancy mm**	**Discrepancy mm**	**Discrepancy mm**	**Discrepancy mm**
Whole cloud point	0.63	0.48	0.72	0.94	0.69 ± 0.33
Menton	0.46	0.24	0.72	0.88	0.57 ± 0.49
Pogonion	0.48	0.32	0.67	0.84	0.58 ± 0.34
B Point	0.44	0.28	0.69	0.81	0.55 ± 0.41
Right Condyle	0.72	0.67	0.81	1.01	0.80 ± 0.26
Left Condyle	0.71	0.63	0.79	1.05	0.79 ± 0.32
Right Gonion	0.55	0.46	0.57	0.92	0.62 ± 0.35
Left Gonion	0.58	0.44	0.54	0.89	0.61 ± 0.34
Right Mental Foramen	0.67	0.35	0.61	0.82	0.61 ± 0.34
Left Mental Foramen	0.64	0.33	0.63	0.79	0.60 ± 0.33
Right Lingula	0.66	0.49	0.66	0.94	0.69 ± 0.32
Left Lingula	0.65	0.41	0.68	0.97	0.68 ± 0.40

4. Discussion

The management of atrophic jaw fractures has always been a topic of discussion in the literature. Luhr et al. [10] in 1996 already classified the atrophic mandibles into three categories, according to the degree of atrophy and the mandibular height: Class I, 16 to 20 mm; Class II, 11 to 15 mm; Class III, <10 mm. In the Class III, bone quality diminished because of a possible sclerotic and because of blood flow decrease. After Luhr et al.'s studies, several authors focused on the correct management and treatment of Class III fractures that often are atrophic pluri-fragmentary mandibular fractures. The problems related to these fractures are associated with patient's age, medical comorbidities, poor bone quality, and decreased vascularity, as well as reduced contact area between the fracture ends [7]. In addition, mandibular atrophy is often the result of complete edentulism: this condition determines the loss of the occlusal reference, which is the guide for the correct reduction of these fractures [6,7,11]. In the literature, the possible treatments for atrophic edentulous mandibular fractures are:

(1) Observation: in case of not dislocated fractures in patients with severe anesthetic risk. The most frequent complication is the non-union of the bone stumps [1];

(2) Closed reduction (mandibulomaxillary fixation—MMF): in case of edentulous atrophic fractures in patients with high anesthetic risk; postoperative malunions and nonunions were very frequent [12];

(3) Gunning splints: the use of gunning splints is preferred in case of edentulous atrophic patients because the open reduction is not helpful due to compromised medical condition of these patients. However, as Dharaskar et al. have shown, there is the possibility of ankylosis, induced by joint blockage (5–6%) [13];

(4) External fixation: it is indicated as a temporary fixation when the patient needs earlier medical treatment. External fixators do not guarantee permanent stability, so malunion and nonunion are common [14];

(5) Open Reduction Internal Fixation (ORIF): with titanium mesh, locking miniplate, or 2.4-reconstruction plate, with or without simultaneous bone grafting. It is indicated for all atrophic surgical mandible fractures. In all cases, the aim of the treatment of edentulous

atrophic mandibular fracture should be to improve patient's quality of life with the minimal risk, ensuring fracture stability [15,16].

Bruce and Ellis (1993) in their review concluded that the optimal treatment for this kind of fractures is an open reduction and stable fixation with large osteosynthesis plates (ORIF) [17]. Different types of techniques were used for this goal such as fixing the bone stumps in occlusion with the patient's dentures [18] or fixing miniplates to the inferior mandibular border to temporarily maintain the fragments alignment before applying a reconstruction plate to vestibular cortical bone [19]. Considering that the use of mini plates can increase the operating times and, furthermore, the patient's dentures are not always available, using a pre-formed plate can become a useful resource both for guiding the reduction in absence of occlusal guide and for reducing intraoperative times. In order to pre-model a 2.4 reconstruction plate, we used Virtual Surgical Planning (VSP). The recent scientific literature has revealed that VSP is a valid tool of assisting surgeons in various procedures, ranging from orthognathic surgery to reconstructive surgery with free flaps, correction of congenital malformations and craniosynostosis, cranio-facial traumatology, distraction osteogenesis, and implantology. Although the use of VSP has been reported in all cited fields, currently there is minimal evidence in elderly patients with atrophic edentulous mandibular fractures [20,21]. In mentioned studies, the authors underlined the importance of VSP to solve the problems related to intraoperative reduction of fragments and to postoperative complications. In fact, without the use of VSP, different complications have been reported: pseudarthrosis of the bone stumps, osteomyelitis, unstable reduction of the fragments that did not allow prosthetic rehabilitation, facial asymmetry, and sensory disturbances of the inferior alveolar nerve [22]. Moreover, the majority of the works in the literature pave the way for the VSP treatment through case reports or case series, but there are no randomized controlled trials on a large sample. [23,24] For this reason, new and extensive contributions to the topic are supported to increase the collective experience in this field. Thus, the aim of our study has been to provide our experience of an in-house digital workflow to show the advantages of VSP in the management and treatment of atrophic edentulous mandibular fractures, where the occlusal guide is absent.

The applied protocol aims to model a reconstruction plate based on the patient's anatomy. Indeed, the time spent for the design phase guarantees a reduction of the operating times. The accuracy of the printing depends on the accuracy of the CT, which must have slices of 1–2 mm. The Formlabs 3B+ 3D print allows to create details with microscopic precision (resolution 25 micrometres); this is a fundamental characteristic for printing instruments and devices in the medical and surgical fields. The thickness of each layer can vary between 25–300 micrometres, for high precision of the anatomical details of the model.

Regarding the surgical procedure, Brucoli et al. [7] in their paper confirmed that, for the treatment of fractures in Luhr's Class III mandibles, the cutaneous external surgical approach is preferred. In fact, this access ensures adequate fragments exposure and, consequently, a more careful surgical procedure. It reduces the risk of contamination by oral microbes and prevents any injury to the inferior alveolar nerve, usually located on top of the alveolar crest in atrophic mandibles [25]. In accordance with the literature, we applied this approach in all patients: thanks to the pre-formed plate, we avoided a cervicotomy to expose the bone fragments, but we used a small sub-mental access. In fact, starting from a minimal sub-mental access, we exposed all the fractures stumps moving the retractors according to the surgical phase.

Recent studies [26] have showed that the use of a pre-formed reconstruction plate can reduce both the time of the manual plate modeling during the surgery and the time for finding the correct position of the plate on the mandibular fragments.

In our cases, we firstly positioned the pre-formed plate on the resin model to calculate the correct position and to estimate the correct screws' number to insert on each mandibular stump. During the surgical procedure, we found that the plate fits perfectly on the bone surfaces, and no further manipulation of the plate was required. Finally, the other fragments were reduced and fixed to the plate as the pre-operative program. At the end of procedure, the surgical time was significantly reduced compared to our standard time, and the accuracy of the treatment was greatly improved.

The success of the treatment was confirmed by the comparison between the virtual surgical planned CT and the post-operative CT. In fact, the overlapping of the two CT images showed that the discrepancies in mm were in all cases less than 1,5 mm. In particular, to verify where the outcome was more predictable, we calculated the discrepancies in eleven bone cephalometric points. The analysis showed that the front portion of mandible (B point and Menton) presented the lower discrepancy, while the higher discrepancy was obtained in the posterior section (condyles and Lingula). This result seems to confirm that the position in the space of the anterior fragments is more easily predictable than the posterior sections. This is probably due to the action/force of the applied plate to the bone that stabilizes the bony portion compared to other areas which are more unstable.

These results were confirmed by the good clinical outcomes in all four cases in terms of facial symmetry restoration, absence of sensory disturbance and possibility of a prosthetic rehabilitation. Furthermore, it is a very low-cost method. If we exclude the initial investment for the purchase of the printer and the dedicated software (around 15,000.00 Euros), the cost for single case is estimated to be around 6.3 Euros. A patient specific plate (PSP) has a commercial cost of around 5000.00 Euros. Therefore, the described in-house workflow allows to amortize initial set-up expenses after a few performed cases.

One of the limitations of the study is certainly the sample size. This is a small sample enrolled in a single center. Further studies or larger case studies with the involvement of other structures are necessary to confirm the results obtained. Other limits are represented by the long learning curve and the time consuming. In fact, the procedure requires specific staff with a background in CAD CAM technologies. This method is reproducible in other clinics by having the medical software and the three-dimensional printer. Another limit of the study is the possible reduction in the plate resistance compared with a custom-made plate, which must not be modeled. However, this reduction of resistance is lower than the conventional method because the plate is subject to less stress, given that it is modeled on a template.

We suggest that future studies could focus on this protocol for other types of fractures, such as comminuted maxillo-mandibular fractures.

5. Conclusions

Within the limits of the study, the Virtual Surgical Planning could be a helpful instrument in the management of the complex fractures in atrophic edentulous mandibles both to reduce the fracture and to decrease the surgical times. The bone repositioning accuracy of our in-house digital workflow is high because all the obtained discrepancy values have been less than 1.5 mm. Nevertheless, the virtual reduction of atrophic mandible fractures is more predictable in the anterior portion.

Author Contributions: Conceptualization, V.A. and U.C.; methodology, S.T.; software, U.C.; validation, G.G., G.D.O. and P.B.; formal analysis, S.T.; investigation, V.A.; resources, P.B.; data curation, F.B. and G.D.O.; writing—original draft preparation, S.T.; writing—review and editing, V.A.; visualization, F.T. and L.A.V.; supervision, L.C.; project administration, L.C.; funding acquisition, L.A.V. All authors have read and agreed to the published version of the manuscript.

Funding: This research received no external funding.

Institutional Review Board Statement: The study was conducted in accordance with the Declaration of Helsinki and was approved by the Ethics Committee of University of Naples Federico II with the protocol number 373/19.

Informed Consent Statement: Informed consent was obtained from all subjects involved in the study. Written informed consent has been obtained from the patients to publish this paper.

Data Availability Statement: Not applicable.

Acknowledgments: Materialize Mimics 21.0 Software.

Conflicts of Interest: The authors declare no conflict of interest.

References

1. Brucoli, M.; Boffano, P.; Romeo, I.; Corio, C.; Benech, A.; Ruslin, M.; Forouzanfar, T.; Rodríguez-Santamarta, T.; Vicente, J.C.; Tarle, M.; et al. The epidemiology of edentulous atrophic mandibular fractures in Europe. *J. Craniomaxillofac. Surg.* **2019**, *47*, 1929–1934. [CrossRef] [PubMed]
2. Clayman, L.; Rossi, E. Fixation of atrophic edentulous mandible fractures by bone plating at the inferior border. *J. Oral Maxillofac. Surg.* **2012**, *70*, 883–889. [CrossRef] [PubMed]
3. Nasser, M.; Fedorowicz, Z.; Ebadifar, A. Management of the fractured edentulous atrophic mandible. *Cochrane Database Syst. Rev.* **2007**, CD006087. [CrossRef]
4. Müller, S.; Bürgers, R.; Ehrenfeld, M.; Gosau, M. Macroplate fixation of fractures of the edentulous atrophic mandible: Immediate function and masticatory rehabilitation. *Clin. Oral Investig.* **2011**, *15*, 151–156.
5. Ravikumar, C.; Bhoj, M. Evaluation of postoperative complications of open reduction and internal fixation in the management of mandibular fractures: A retrospective study. *Indian J. Dent. Res.* **2019**, *30*, 94–96. [PubMed]
6. Florentino, V.G.B.; Abreu, D.F.; Ribeiro, N.R.B.; Silva, L.F.; Gondin, R.F.; Mello, M.J.R.; Aguiar, A.S.W. Surgical treatment of bilateral atrophic mandible fracture. *J. Craniofac. Surg.* **2020**, *31*, e753–e755. [CrossRef]
7. Brucoli, M.; Boffano, P.; Romeo, I.; Corio, C.; Benech, A.; Ruslin, M.; Forouzanfar, T.; Rodríguez-Santamarta, T.; Vicente, J.C.; Tarle, M.; et al. Surgical management of unilateral body fractures of the edentulous atrophic mandible. *Oral Maxillofac. Surg.* **2020**, *24*, 65–71. [CrossRef]
8. Nilsson, J.; Nysjö, F.; Nyström, I.; Kampe, J.; Thor, A. Evaluation of in-house, haptic assisted surgical planning for virtual reduction of complex mandibular fractures. *Int. J. CARS* **2021**, *16*, 1059–1068. [CrossRef]
9. Friscia, M.; Seidita, F.; Committeri, U.; Troise, S.; Abbate, V.; Bonavolontà, P.; Dell'Aversana Orabona, G.; Califano, L. Efficacy of Hilotherapy face mask in improving the trend of edema after orthognathic surgery: A 3D analysis of the face using a facial scan app for iPhone. *Oral Maxillofac. Surg.* **2022**, *26*, 485–490. [CrossRef]
10. Luhr, H.G.; Reidick, T.; Merten, H.A. Frakturen des atrophischen Unterkiefers—eine Herausforderung für die Therapie [Fractures of the atrophic mandible—a challenge for therapy]. *Fortschr. Kiefer Gesichtschir.* **1996**, *41*, 151–154. (In German)
11. Dell' Aversana Orabona, G.; Iaconetta, G.; Abbate, V.; Califano, L. Bifocal mandibular fractures: Which should be treated first? *J. Craniofac. Surg.* **2012**, *23*, 1723–1727. [CrossRef]
12. De Feudis, F.; De Benedittis, M.; Antonicelli, V.; Pittore, P.; Cortelazzi, R. Decision-making algorithm in treatment of the atrophic mandible fractures. *G. Chir.* **2014**, *35*, 94–100. [PubMed]
13. Dharaskar, S.; Athavale, S.; Kakade, D. Use of gunning splint for the treatment of edentulous mandibular fracture: A case report. *J. Indian Prosthodont. Soc.* **2013**, *14*, 415–418. [CrossRef] [PubMed]
14. Kazi, A.A.; Lee, T.S.; Vincent, A.; Sokoya, M.; Sheen, D.; Ducic, Y. The role of external fixation in trauma and reconstruction of the mandible in the age of rigid fixation. *Facial Plast. Surg.* **2019**, *35*, 614–622. [CrossRef]
15. Wittwer, G.; Adeyemo, W.L.; Turhani, D.; Ploder, O. Treatment of atrophic mandibular fractures based on the degree of atrophy—experience with different plating systems: A retrospective study. *J. Oral Maxillofac. Surg.* **2006**, *64*, 230–234. [CrossRef]
16. Ellis, E., 3rd; Price, C. Treatment protocol for fractures of the atrophic mandible. *J. Oral Maxillofac. Surg.* **2008**, *66*, 421–435. [CrossRef]
17. Bruce, R.A.; Ellis, E. The second Chalmers J. Lyons academy study of fractures of the edentulous mandible. *J. Oral Maxillofac. Surg.* **1993**, *51*, 904–911. [CrossRef] [PubMed]
18. Agochukwu, N.B.; Maus, J.; Wang, D.; Stewart, D. Use of the MatrixWAVE™ system with dentures to establish maxillomandibular fixation in edentulous patients. *Br. J. Oral Maxillofac. Surg.* **2018**, *56*, 343–345. [CrossRef] [PubMed]
19. Franciosi, E.; Mazzaro, E.; Larranaga, J.; Rios, A.; Picco, P.; Figari, M. Treatment of edentulous mandibular fractures with rigid internal fixation: Case series and literature review. *Craniomaxillofac. Trauma Reconstr.* **2014**, *7*, 35–42. [CrossRef]
20. Kupfer, P.; Saadad, N.; Hughes, P.J. Open reduction and internal fixation of bilateral atrophic mandible fractures utilizing virtual surgical planning, custom cutting guides and reconstruction plate. A case report. *J. Oral Maxillofac. Surg* **2016**, *74*, e89. [CrossRef]
21. Broyles, J.M.; Wallner, C.; Borsuk, D.E.; Dorafshar, A.H. The role of computer-assisted design and modeling in an edentulous mandibular malunion reconstruction. *J. Craniofac. Surg.* **2013**, *24*, 1835–1838. [CrossRef] [PubMed]
22. Luhr, H.G.; Reidick, T.; Merten, H.A. Results of treatment of fractures of the atrophic edentulous mandible by compression plating: A retrospective evaluation of 84 consecutive cases. *J. Oral Maxillofac. Surg.* **1996**, *54*, 250–254. [CrossRef] [PubMed]
23. Façanha de Carvalho, E.; Alkmin Paiva, G.L.; Yonezaki, F.; Machado, G.G. Computer-aided surgical simulation in severe atrophic mandibular fractures: A new method for guided reduction and temporary stabilization before fixation. *J. Oral Maxillofac. Surg.* **2021**, *79*, 892.e1–892.e7. [CrossRef] [PubMed]
24. Maloney, K.D.; Rutner, T. Virtual surgical planning and hardware fabrication prior to open reduction and internal fixation of atrophic edentulous mandible fractures. *Craniomaxillofac. Trauma Reconstr.* **2019**, *12*, 156–162. [CrossRef]

25. Madsen, M.J.; Haug, R.H.; Christensen, B.S.; Aldridge, E. Management of atrophic mandible fractures. *Oral Maxillofac. Surg. Clin. N. Am.* **2009**, *21*, 175–183. [CrossRef]
26. Huang, Y.; Xia, Z.; Zhang, X.; Liao, X.; Guo, Z.; Ji, S.; Long, J. Combined use of specially-designed digital surgical guides and pre-formed reconstruction plate to treat bilateral mandibular fracture. *J. Craniofac. Surg.* **2019**, *30*, 2253–2256. [CrossRef]

applied sciences

MDPI

Article

Effectiveness of Machine Learning in Assessing the Diagnostic Quality of Bitewing Radiographs

Mohammed A. Barayan [1], Arwa A. Qawas [2], Asma S. Alghamdi [3], Turki S. Alkhallagi [4], Raghad A. Al-Dabbagh [4], Ghadah A. Aldabbagh [5,6] and Amal I. Linjawi [2,*]

1 Department of Diagnostic Oral Sciences, Faculty of Dentistry, King Abdulaziz University, Jeddah 22254, Saudi Arabia
2 Department of Orthodontics, Faculty of Dentistry, King Abdulaziz University, Jeddah 21589, Saudi Arabia
3 Department of Restorative Dentistry, Faculty of Dentistry, King Abdulaziz University, Jeddah 21589, Saudi Arabia
4 Department of Oral and Maxillofacial Prosthodontics, Faculty of Dentistry, King Abdulaziz University, Jeddah 21589, Saudi Arabia
5 Department of Computer Science, Faculty of Computing and Information Technology, King Abdulaziz University, Jeddah 21589, Saudi Arabia
6 Deparmtne of Mechanical Engineering, Massachusetts Institute of Technology, Cambridge, MA 02139, USA
* Correspondence: ailinjawi@kau.edu.sa; Tel.: +966-504-155573

Featured Application: Machine learning is becoming one of the major platforms for advances in all fields of dentistry. Their further exploration in automating the diagnostic process is of great importance to the field. The aim of this study is to assess the effectiveness of machine learning (ML) in assessing the diagnostic quality of bitewing (BW) radiographs at contact areas between teeth, which can help the oral radiologists in providing better radiographic qualities.

Abstract: Background: Identifying the diagnostic value of bitewing radiographs (BW) is highly dependent on the operator's knowledge and experience. The aim of this study is to assess the effectiveness of machine learning (ML) to classify the BW according to their diagnostic quality. Methods: 864 BW radiographs from records of 100 patients presented at King Abdulaziz University Dental Hospital, Jeddah, Saudi Arabia were assessed. The radiographic errors in representing proximal contact areas (n = 1951) were categorized into diagnostic and non-diagnostic. Labeling and training of the BW were done using Roboflow. Data were divided into validation, training, and testing sets to train the pre-trained model Efficientdet-d0 using TensorFlow. The model's performance was assessed by calculating recall, precision, F1 score, and log loss value. Results: The model excelled at detecting "overlap within enamel" and "overlap within restoration (clear margins) with F1 score of 0.89 and 0.76, respectively. The overall system errors made by the built model showed a log loss value of 0.15 indicating high accuracy of the model. Conclusions: The model is a "proof of concept" for the effectiveness of ML in diagnosing the quality of the BW radiographs based on the contact areas. More dataset specification and optimization are needed to overcome the class imbalance.

Keywords: artificial intelligence; bitewing radiographs; contact areas; machine learning; radiographic errors

Citation: Barayan, M.A.; Qawas, A.A.; Alghamdi, A.S.; Alkhallagi, T.S.; Al-Dabbagh, R.A.; Aldabbagh, G.A.; Linjawi, A.I. Effectiveness of Machine Learning in Assessing the Diagnostic Quality of Bitewing Radiographs. *Appl. Sci.* **2022**, *12*, 9588. https://doi.org/10.3390/app12199588

Academic Editors: Giuseppe Minervini and Rocco Franco

Received: 13 August 2022
Accepted: 21 September 2022
Published: 24 September 2022

1. Introduction

Bitewing radiographs are a vital adjunct to clinical examination for diagnosis and treatment planning of dental and supporting tissue diseases [1]. A balance between risks and benefits of acquiring radiographs, following standards of justification for radiation dose protection, optimizing, and restricting radiation dose, should be achieved [2]. Reducing the radiation dose could be partly achieved by ensuring the acquisition of diagnostically acceptable radiographic images from the first attempt. However, bitewing radiographs are sometimes re-taken for diagnostic purposes. Indeed, Yeung et al., 2021, demonstrated in a systematic review that the rejection rate of bitewings because of its limited diagnostic value was 9%. The most common reasons for rejection were positional errors and cone cuts. It was further suggested that reject rates were higher with digital imagining, specifically when using intraoral sensors [3].

The decision-making process involved in identifying the diagnostic value of bitewing images is complex and is highly dependent on the operator's/technician's knowledge and experience. In general, bitewing images should equally capture both maxillary and mandibular crowns and crestal bone. The premolar bitewings should capture the distal surfaces of both maxillary and mandibular canines, while molar bitewings should capture the distal surface of the second premolar and the distal surface of the last standing molar in the mouth. Additionally, there should be no overlap of proximal surfaces in contact areas [4]. One possible adjunct that may improve this decision-making process could be the use of Artificial Intelligence (AI) programs in the field.

AI refers to any computer or technology capable of mimicking human thought processes such as problem-solving. Additionally, it can refer to the capacity of computers that can learn from data and represent their own intelligence so that issues can be solved from this learning [5]. This science was first developed by John McCarthy, who established the first AI laboratory in 1957 [6]. AI varies in types and classifications depending on its complexity in performance. Machine learning (ML) is a type of AI that makes software application more accurate in predicting outcomes without programming. Deep Learning (DL) is another type of AI defined as "computational systems that learn over time based on experience" [7–10].

Both ML and DL are considered as implementations of computational methods (i.e., algorithms) through which models can be generated from any given data, and those models can then be used for further prediction of new data sets with similar properties [11]. The differentiation between ML and DL can be made based on the complexity of the network used to extract features from the data. It is also related to the complexity of the data itself. ML is usually appropriate for regression or classification of simple numerical data. DL is recommended for the analysis of large collections of images or other complex data [8–11].

Clinical decision–support systems (CDSS) are software-based AI programs. They can be based on either ML or so-called rules-based expert systems [7,12]. CDSS are designed to provide the health care providers with specialist assistance in their daily activities, helping with tasks that depend on data and information exploitation [7]. There are many types of CDSS systems, such as artificial neural networks (ANNs), fuzzy expert systems, evolutionary computation, and hybrid intelligent systems. They vary in methods of handling data and the complexity of the analysis process [7,10].

Nowadays, the application of AI is increasing in multiple dental and medical applications with promising results [7,10]. Radiographic images are considered an important source of data for developing AI models in the health professional fields. Therefore, it is important to consider the factors in the quality of radiographic images that can affect the appearance of structures [7]. AI application in dentistry was assessed for three main purposes: (1) as a triage tool for optimizing the clinical workflow, (2) as a diagnostic tool for the separation of normal from abnormal conditions, and (3) as a treatment decision-making tool for the detection of specific diseases [7].

Studies in the literature about the value of AI from bitewing radiographs in assisting dentists in decision-making are plentiful. In general, these studies demonstrate the utility of AI algorithms in diagnosing dental caries, periodontal disease, and classifying restorations [13–15]. Most of these studies investigated the effectiveness of AI models from bitewings in diagnosing dental caries. These studies demonstrated that deep learning-based AI algorithms increased the accuracy of diagnosing proximal dental caries [13,16–18]. A few studies investigated the utility of AI from bitewings in diagnosing periodontal disease. Overall, AI models for diagnosing periodontitis are still developing but might be useful adjunctive tools [16].

In one study, Mertens et al., 2021, demonstrated that deep learning-based CNNs from periapical and bitewing radiographs was a potentially beneficial technique in detecting and differentiating restorations [14]. However, there are no studies in the literature that investigate the development or utility of AI algorithms to classify bitewings based on their diagnostic value. Accordingly, the aim of this study is to develop and assess the effectiveness of machine learning-based AI algorithm to classify bitewing radiographs according to their diagnostic quality.

2. Materials and Methods

2.1. Study Design

This is a laboratory study approved by the ethical committee at King Abdulaziz University, Faculty of Dentistry, Jeddah, Saudi Arabia (Ethical no.: 172-12-20). The study was conducted in collaboration with the Faculty of Computing and Information Technology at the same institute.

2.2. Data Collection

A research account was created for the investigators to access the patients' records through the University Dental Hospital health record system (CareStream R4 Clinical+ software, Jeddah, Saudi Arabia), and to retrieve the required BW radiographs. The inclusion criteria for the BW radiographs were: (1) for patients under treatment at the University Dental Hospital with fully erupted permanent dentition, and (2) the BW was either premolar or molar view. The exclusion criteria included: (1) records with findings that might affect the machine learning process such as severe periodontitis that cannot be assessed by a horizontal BW, removable partial dentures components, fixed retainers, provisional crowns, and defective direct/indirect restorations that hinder the contacts assessment, and (2) certain acquisition errors such as double exposure, patient movement, or extremely low/high kVp that resulted in non-distinguishable enamel, dentin, or pulp, and cannot be corrected. The R4 Clinical+ system was searched for patients' records from 2017–2021 that fulfilled the study inclusion criteria.

2.3. Data Preprocessing and Entry

Two investigators, who were general dental practitioners, were trained and calibrated to perform labeling under the supervision of an experienced Oral and Maxillofacial Radiologist.

The selected BW radiographs were reviewed and had the rotation and contrast manually adjusted, if needed, to an acceptable level of differentiation between enamel, dentin, pulp, and air; to aid the model in the machine learning process, then exported as TIFF format with dimensions of 1920 × 1440 pixels.

For each patient, two categories of data were collected: demographic data and radiographic data. The demographic data included: patient's file number, date of birth, nationality, ethnicity, and number of bitewings taken.

The radiographic data included the following datasets:

1. Radiographic ID: Each BW image was renamed with ten digits. The last seven digits referred to the patient's file number on the R4 Clinical+ software, while the first three numbers represented the BW radiograph number for each specific patient.
2. Acquisition date: the date when the BW radiograph was taken in MM/DD/YYYY format.
3. Machine type: the type of machine used to acquire the BW radiographs, either sensor or digitalized plates.
4. Radiograph view side: which side of the dentition was included in the BW radiograph (right or left).
5. Score: the scoring system for each proximal surface and contact area between two teeth was established based on tooth structure, amount of contact overlap, and the field of the taken radiograph. The scores ranged from 0 to 11.

2.4. Dataset

After scoring all selected bitewing radiographs, feature selection was performed to identify prominent features leading to successful classification results. Accordingly, the labels that were included in the model training process were the following:

1. Overlap within enamel.
2. Overlap within a restoration (clear margin).
3. Overlap at DEJ.
4. Overlap within dentin.
5. Overlap within a restoration (unclear margin).

Since the machine learning task is an object detection task, the "out-of-field" label had to be excluded, as the contacts in this case are not present in the images and thus cannot be identified using bounding boxes. In addition, labels that were not errors, such as missing tooth/surface, and open contacts have also been excluded to avoid misclassification.

Images were converted from TIFF to JPG format then manually labeled using Roboflow, a tool created by for-profit Roboflow company that facilitates collaborative image annotation and allows for easy split, augmentation, and exportation. The resulting dataset was then split into three sets: training set (583 images), validation set (167 images), and test set (84 images).

It seems that since the data were randomly collected, and some labels of more common errors are over-represented in the dataset, as shown in Table 1. Hence, the model would probably excel at detecting these labels more than it would with others. On the other hand, we can see that "overlap with restoration -unclear margin-", "overlap within dentin", "overlap at DEJ", are all under-represented, thus it is expected from the model to not be as efficient at detecting these classes.

Table 1. Contact areas label distribution.

Class	Contact Area Count (%) in the Whole Dataset	Contact Area Count (%) in the Training Set
Overlap within enamel	1124 (57.6)	555 (52.8)
Overlap within restoration (clear margins)	352 (18.0)	226 (21.5)
Overlap at DEJ	162 (8.3)	83 (7.9)
Overlap within dentin	109 (5.6)	64 (6.1)
Overlap within restoration (unclear margins)	204 (10.5)	123 (11.7)
Total	1951 (100)	1051 (100)

To overcome the imbalance issue, oversampling and under sampling techniques were applied for over-represented labels, and randomly selected images were removed. To increase under-represented labels image augmentation was performed, where we increased the number of images that contained these labels by using shift, flip, brightness, and zoom image data augmentation.

2.5. Model Training

The model was created using TensorFlow Object Detection API "Application Programming Interface", an open-source platform with a preceding software library that was developed by Google Brain team. It allows the use and customization of many saved networks termed as pre-trained models. They were previously trained on a large general dataset, which meant that a pre-trained model could be taken advantage of to effectively work as a learned generic model, which aids in reducing the amount of data needed to train a model from scratch [19].

Several pre-trained models were experimented with. These models are: ssd_mobilenet_v2, RetinaNet, and EfficientDet-D0, which were all pre-trained on coco_2018 dataset.

Among the three different object detection model architectures, EfficientDet achieved the best performance, as it required the lowest number of training epochs with similar accuracy. Accordingly, the EfficientDet-D0 model was chosen for this study, which is a light and small version of EfficientDet (Figure 1) [20].

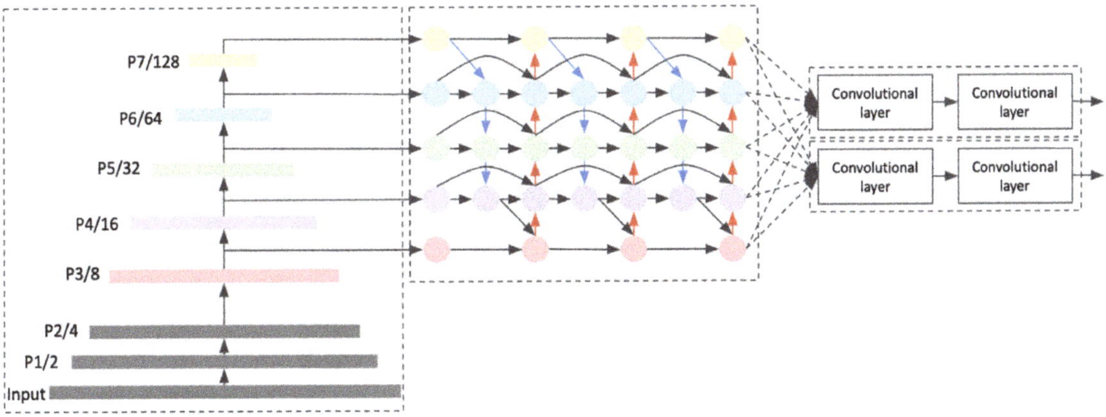

Figure 1. EfficientDet architecture that consists of EfficientNet backbone, BiFPN as feature network neck, and output head.

In this study, a fine-tuning approach was used for pre-trained model customization. This is done by unfreezing then training the last layers of the frozen model base, together with the newly added classifier layers. This allows us to modify the representations needed in the model for feature detection and make them more relevant for the specified task [19]. Training of the model to identify errors was done using the training set within an open-source product from Google Research, called Google Colab, using GPU "Graphics Processing Unit" runtime. The initial learning rate was chosen as 0.001 with further exponential decay. There were 7000 training steps, with a batch size of 16 [21].

2.6. Model Assessment

To evaluate inference performance, each radiograph was compared to the ground truth labeling provided previously, using the data of the testing set that was not seen by the system during the training phase.

2.7. Statistical Analysis

Cohen's kappa coefficient was calculated to assess intra-rater and inter-rated reliability. The performance of the machine learning model was evaluated by the following classification metrics.

1. Log loss: represents the summation of system errors.
2. Recall: quantifies the number of correct positive predictions made from all positive predictions that could have been made.
3. Precision: quantifies the number of correct positive predictions made by the model.
4. F1 score: combines recall and precision into a single score by calculating the harmonic mean of precision and recall, which is the measure that is used to evaluate this model's performance.

3. Results

3.1. Data Analysis

Cohen's kappa coefficient for intra-rater and inter-rater reliability were 0.95 and 0.96, respectively, indicating excellent agreement.

A total of 834 randomly chosen BW radiographs of 100 patients presented to King Abdulaziz University Dental Hospital were collected with a minimum number of one BW and a maximum of 33 BW per patient (mean $\pm$ SD; 15.08 $\pm$ 9.553). The final dataset consisted of 583 images for the training set, 167 for the validation set, and 84 for the test set. Details of numbers of labeled contact area within the assessed dataset and training set radiographs are presented in Table 1.

Results revealed class imbalance, which means over-representation of some labels, such as for "overlap within enamel", while others were under-represented, such as "overlap at DEJ" and "overlap within dentin" (Table 1, Figure 2).

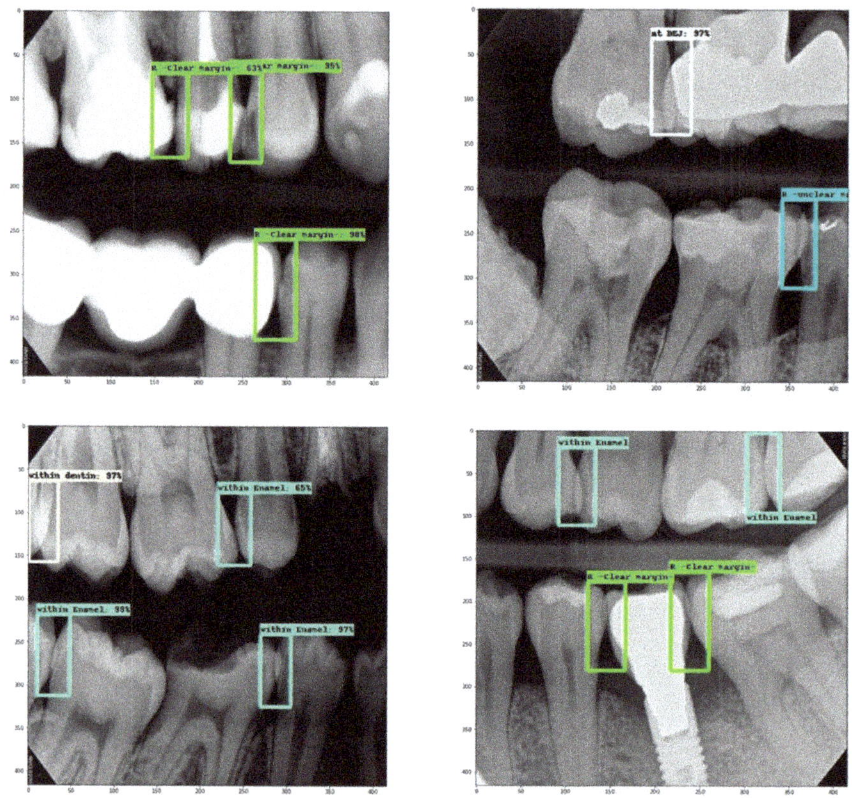

Figure 2. Representative samples of assessed contact area overlap; Cyan box: within enamel, White box: at DEJ, Off-white box: within dentin, Green box: within restoration (clear margins), Blue box: within restoration (unclear margins).

3.2. Machine Learning Performance Assessment

To obtain evaluation metrics, a confusion matrix was constructed for each error label (X):

- "True positives (TP) of "X" are all X instances that are classified as X;
- "True negatives (TN) of "X" are all non-X instances that are not classified as X;
- "False positives (FP) of "X" are all non-X instances that are classified as X;
- "False negatives (FN) of "X" are all X instances that are not classified as X.

True negative (TN) metric is not useful for object detection. Therefore, it was ignored. The following metrics were calculated as seen in Table 2 using the confusion matrix: recall (sensitivity) = TP/TP + FN, precision = TP/TP + FP, and F1 score = (2 $\times$ Precision $\times$ Recall)/(Precision + Recall), where TP, FP, FN represent true-positive, false-positive, and false-negative results, respectively.

Table 2. Precision, recall, and F1 score per class.

	Overlap within Enamel	Overlap within Restoration (Clear Margin)	Overlap at DEJ	Overlap within Dentin	Overlap within Restoration (Unclear Margin)
TP	133	13	7	3	2
FP	16	0	5	2	0
FN	15	8	5	11	4
Precision	0.893	(1no FP)	0.583	0.600	(1no FP)
Recall	0.899	0.619	0.583	0.214	0.333
F1 score	0.896	0.764	0.583	0.316	0.499

Results showed that the model excelled in detecting the following two classes: "overlap within enamel" (Precision: 0.893, Recall: 0.899, F1 score: 0.896) and "overlap within restoration "clear margins)" (Precision: 0.893, Recall: 0.619, F1 score: 0.764) (Table 2). However, for the other classes, the evaluation metrics were unreliable due to under-representation of those classes (Table 2). The overall system errors made by this built model was represented by the log loss value, which had a value of 0.15.

3.3. Web App

As an additional step, a preliminary trial of building a web app was done to check the compatibility with the system used at King Abdulaziz University Dental Hospital. The app was built using Html and CSS for front-end, Flask framework for back-end, and Google App Engine for hosting. App Engine was used because it allows for creating custom deployment environments which is necessary for such applications (34.122.82.96, the link is protected by a password). The app was then tested on a number of images and showed promising results in detecting the classes at contact areas, as shown in Figure 3. However, this was not part of the main objective of the study, and thus will need further assessment on a larger sample size.

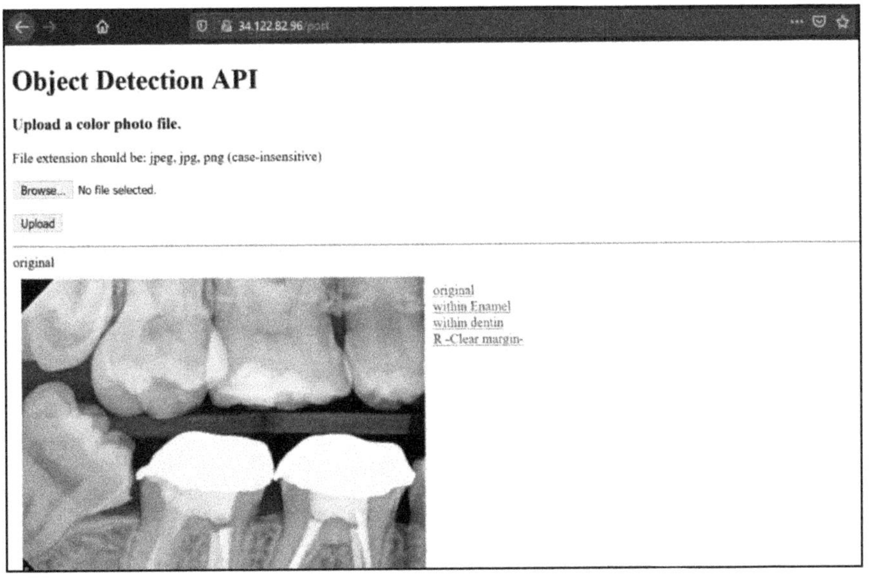

Figure 3. The model's interface (an unseen radiograph was uploaded through the interface and the model detected three errors of different classes: within enamel, within dentin, and within restoration (clear margins)).

4. Discussion

The aim of the current study was to assess the effectiveness of ML in detecting the diagnostic quality of BW radiographs so it can be used as a decision–support tool to help technicians check the quality of the taken radiographs faster and with less effort [6,16]. The diagnostic quality of the BW depends on many factors and features, however, this study focused on the assessment of the contact areas as an important feature of the quality of the BW images. The detection of different radiographic errors regarding the contact area is subject to the discriminatory skills of examiners which varies significantly and is heavily related to examiner experience. Therefore, the application of ML to detect radiographic errors will aid in reducing this variation significantly.

Scoring proximal contact areas were developed through stages—from a comprehensive list to a more compact encompassing one—in this project. The objective was to identify prominent features in bitewing radiographs for a successful classification of diagnostic and non-diagnostic images and to yield a less complex model for the AI program. Initially, we identified eleven possible proximal contact area presentations based on tooth structure, amount of proximal contact overlap, and the acquired radiographic field. Then, to facilitate cross-tabulation and coding of proximal contact areas, these scores were further grouped in terms of diagnostic value into four basic groups. Finally, images were labeled based on the most prominent proximal contact area features to aid in identifying the diagnostic value. This included proximal contact overlap within tooth (enamel, DEJ, and dentin) or within restoration (with or without clear margins).

In this study, Python and Python-friendly environments were used to build the model, which is a popular and powerful interpreted IT language. The descriptive analysis of the dataset revealed class imbalance, which means over-representation of some labels while others were under-represented. The class "overlap within enamel" was over-represented and the class "overlap within restoration (clear margin) was adequately represented in the dataset, while the classes "overlap within restoration (unclear margin), "overlap within dentin", and "overlap at DEJ" were all under-represented in the dataset. Thus, it is expected that the model will be more efficient at detecting the well-represented classes compared to the under-represented classes.

Additionally, most of the assessed proximal contact areas, if they had errors, were still of diagnostic value (overlap within enamel or within restoration with clear margin). This could partly be because all assessed radiographs in this study were taken with the holder technique. Indeed, it was reported that bitewing radiographs that were acquired by the holder technique (Rinn XCP film positioning devices) had fewer horizontal errors than those that were taken by the loop technique (conventional method) [22].

As for ML performance analysis, recall, precision, and F1 score are all highly affected by the class imbalance and thus are not suitable evaluation metrics for the current model. Thus, log loss was the best evaluation metric to indicate our model accuracy. The loss is calculated on training and validation and its interpretation is how well the model is doing for these two sets. Unlike other accuracy metrics, loss is not a percentage. It is a summation of the errors made for each example in training or validation sets. The lower the loss, the better the accuracy of the model. Log loss nearer to 0 indicates high accuracy, whereas if the log loss was away from 0 then it indicates low accuracy. The built model in the current study had a training loss of 0.15, which indicates high accuracy.

The current findings have several impacts in the field of dentistry. The availability of a machine learning model to assess the quality of radiographs will reduce the errors of misreading the radiographs from both the unskilled technicians and dentists. It will also help the radiology department in a busy dental practice to have efficient outcomes by reading large number of radiographs more efficiently and in less time.

This study has several limitations. For instance, the resulting model is a "proof of concept" model; it is not by any means ready for production. Many different adaptations, tests, and experiments still need to be conducted. The model is yet to be deployed and integrated with the web-app. However, to enhance the accuracy of the model, class imbalance needs to be managed by collecting more label specific data instead of random data. Despite such limitation, the model has proven that ML could be used to help identify errors within contacts in BW radiographs, with the ability to be continuously reinforced by new data to increase the precision level. The results of the current study add to other findings in the literature about the promising efficacy of AI and ML application in dentistry [5,7,12,23–28].

5. Conclusions

The model is a "proof of concept" for the effectiveness of ML in diagnosing the quality of the BW radiographs based on the contact areas. More dataset specification, optimization, and reinforced learning are needed to overcome the class imbalance.

Author Contributions: Conceptualization, G.A.A., R.A.A.-D., A.I.L., T.S.A. and M.A.B.; methodology, G.A.A., R.A.A.-D., A.I.L., T.S.A. and M.A.B.; software, G.A.A., R.A.A.-D., A.I.L., T.S.A., M.A.B., A.A.Q. and A.S.A.; validation, G.A.A., R.A.A.-D., A.I.L., T.S.A., M.A.B., A.A.Q. and A.S.A.; formal analysis, G.A.A., R.A.A.-D., A.I.L., T.S.A., M.A.B., A.A.Q. and A.S.A.; investigation, G.A.A., R.A.A.-D., A.I.L., T.S.A., M.A.B., A.A.Q. and A.S.A.; resources, R.A.A.-D., A.I.L., T.S.A., M.A.B., A.A.Q. and A.S.A.; data curation, G.A.A., R.A.A.-D., A.I.L., T.S.A., M.A.B., A.A.Q. and A.S.A.; writing—original draft preparation, R.A.A.-D., A.I.L., T.S.A., M.A.B., A.A.Q. and A.S.A.; writing—review and editing, G.A.A., R.A.A.-D., A.I.L., T.S.A., M.A.B., A.A.Q. and A.S.A.; visualization, G.A.A., R.A.A.-D., A.I.L., T.S.A. and M.A.B.; supervision, G.A.A., R.A.A.-D., A.I.L., T.S.A. and M.A.B.; project administration, G.A.A., R.A.A.-D. and A.I.L. All authors have read and agreed to the published version of the manuscript.

Funding: This research received no external funding.

Institutional Review Board Statement: All data were evaluated retrospectively, and ethical approval was taken from the Ethical Committee at King Abdulaziz University, Faculty of Dentistry, Jeddah, Saudi Arabia [Ethical no.: 172-12-20]. Formal consent is not required for this type of study.

Informed Consent Statement: All data were evaluated retrospectively, thus, informed consent statement is not applicable.

Data Availability Statement: Data supporting reported results will be available on request.

Acknowledgments: We would like to thank Maryam Omer and Hind Tayeb from the Faculty of Computing and Information Technology, King Abdulaziz University, Jeddah, Saudi Arabia for their great support and contribution in running the machine learning model and running the tests.

Conflicts of Interest: The authors declare no conflict of interest.

References

1. Boeddinghaus, R.; Whyte, A. Trends in maxillofacial imaging. *Clin. Radiol.* **2018**, *73*, 4–18. [CrossRef] [PubMed]
2. Jaju, P.P.; Jaju, S.P. Cone-beam computed tomography: Time to move from ALARA to ALADA. *Imaging Sci. Dent.* **2015**, *45*, 263–265. [CrossRef] [PubMed]
3. Yeung, A.W.; Wong, N.S. Reject Rates of Radiographic Images in Dentomaxillofacial Radiology: A Literature Review. *Int. J. Environ. Res. Public Health* **2021**, *18*, 8076. [CrossRef]
4. White, S.C.; Pharoah, M.J. Intraoral projections. In *Oral Radiology–Principles and Interpretation*, 7th ed.; Elsevier Mosby: Philadelphia, PA, USA, 2014; pp. 91–130.
5. Khanagar, S.B.; Al-Ehaideb, A.; Maganur, P.C.; Vishwanathaiah, S.; Patil, S.; Baeshen, H.A.; Sarode, S.C.; Bhandi, S. Developments, application, and performance of artificial intelligence in dentistry–A systematic review. *J. Dent. Sci.* **2021**, *16*, 508–522. [CrossRef] [PubMed]
6. Rajaraman, V. JohnMcCarthy—Father of artificial intelligence. *Resonance* **2014**, *19*, 198–207. [CrossRef]
7. Yaji, A.; Prasad, S.; Pai, A. Artificial intelligence in dento-maxillofacial radiology. *Acta Sci. Dent. Sci.* **2019**, *3*, 116–121.
8. Heidari, A.; Jafari Navimipour, N.; Unal, M.; Toumaj, S. Machine learning applications for COVID-19 outbreak management. *Neural Comput. Appl.* **2022**, *10*, 15313–15348. [CrossRef]
9. Heidari, A.; Toumaj, S.; Navimipour, N.J.; Unal, M. A privacy-aware method for COVID-19 detection in chest CT images using lightweight deep conventional neural network and blockchain. *Comput. Biol. Med.* **2022**, *145*, 105461. [CrossRef]
10. Heidari, A.; Navimipour, N.J.; Unal, M.; Toumaj, S. The COVID-19 epidemic analysis and diagnosis using deep learning: A systematic literature review and future directions. *Comput. Biol. Med.* **2021**, *14*, 105141. [CrossRef]
11. Pauwels, R. A brief introduction to concepts and applications of artificial intelligence in dental imaging. *Oral Radiol.* **2021**, *37*, 153–160. [CrossRef]

12. Ekert, T.; Krois, J.; Meinhold, L.; Elhennawy, K.; Emara, R.; Golla, T.; Schwendicke, F. Deep learning for the radiographic detection of apical lesions. *J. Endod.* **2019**, *45*, 917–922. [CrossRef] [PubMed]
13. Mohammad-Rahimi, H.; Motamedian, S.R.; Rohban, M.H.; Krois, J.; Uribe, S.; Nia, E.M.; Rokhshad, R.; Nadimi, M.; Schwendicke, F. Deep learning for caries detection: A systematic review: DL for Caries Detection. *J. Dent.* **2022**, *30*, 104115. [CrossRef]
14. Karatas, O.; Cakir, N.N.; Ozsariyildiz, S.S.; Kis, H.C.; Demirbuga, S.; Gurgan, C.A. A deep learning approach to dental restoration classification from bitewing and periapical radiographs. *Quintessence Int.* **2021**, *52*, 568–574. [PubMed]
15. Revilla-León, M.; Gómez-Polo, M.; Barmak, A.B.; Inam, W.; Kan, J.Y.; Kois, J.C.; Akal, O. Artificial intelligence models for diagnosing gingivitis and periodontal disease: A systematic review. *J. Prosthet. Dent.* **2022**. *online ahead of print*. [CrossRef]
16. Bayrakdar, I.S.; Orhan, K.; Akarsu, S.; Çelik, Ö.; Atasoy, S.; Pekince, A.; Yasa, Y.; Bilgir, E.; Sağlam, H.; Aslan, A.F.; et al. Deep-learning approach for caries detection and segmentation on dental bitewing radiographs. *Oral Radiol.* **2021**, *22*, 468–479. [CrossRef] [PubMed]
17. Mertens, S.; Krois, J.; Cantu, A.G.; Arsiwala, L.T.; Schwendicke, F. Artificial intelligence for caries detection: Randomized trial. *J. Dent.* **2021**, *115*, 103849. [CrossRef] [PubMed]
18. Devlin, H.; Williams, T.; Graham, J.; Ashley, M. The ADEPT study: A comparative study of dentists' ability to detect enamel-only proximal caries in bitewing radiographs with and without the use of AssistDent artificial intelligence software. *Br. Dent. J.* **2021**, *231*, 481–485. [CrossRef]
19. TensorFlow. Transfer Learning and Fine-Tuning | TensorFlow Core. 2021. Available online: https://www.tensorflow.org/ (accessed on 1 November 2021).
20. Tan, M.; Pang, R.; Le, Q.V. Efficientdet: Scalable and efficient object detection. In Proceedings of the IEEE/CVF Conference on Computer Vision and Pattern Recognition, Seattle, WA, USA, 13–19 June 2020; pp. 10781–10790.
21. Google Colab Notebook. Available online: https://colab.research.google.com (accessed on 1 November 2021).
22. Potter, B.J.; Shrout, M.K.; Harrell, J.C. Reproducibility of beam alignment using different bite-wing radiographic techniques. *Oral Surg. Oral Med. Oral Pathol. Oral Radiol. Endod.* **1995**, *79*, 532–535. [CrossRef]
23. Saghiri, M.A.; Garcia-Godoy, F.; Gutmann, J.L.; Lotfi, M.; Asgar, K. The reliability of artificial neural network in locating minor apical foramen: A cadaver study. *J. Endod.* **2012**, *38*, 1130–1134. [CrossRef]
24. Devito, K.L.; de Souza Barbosa, F.; Felippe Filho, W.N. An artificial multilayer perceptron neural network for diagnosis of proximal dental caries. *Oral Surg. Oral Med. Oral Pathol. Oral Radiol.* **2008**, *106*, 879–884. [CrossRef]
25. Johari, M.; Esmaeili, F.; Andalib, A.; Garjani, S.; Saberkari, H. Detection of vertical root fractures in intact and endodontically treated premolar teeth by designing a probabilistic neural network: An ex vivo study. *Dentomaxillofac. Radiol.* **2017**, *46*, 20160107. [CrossRef] [PubMed]
26. Lee, J.H.; Kim, D.H.; Jeong, S.N.; Choi, S.H. Diagnosis and prediction of periodontally compromised teeth using a deep learning-based convolutional neural network algorithm. *J. Periodontal. Implant. Sci.* **2018**, *48*, 114–123. [CrossRef] [PubMed]
27. Aubreville, M.; Knipfer, C.; Oetter, N.; Jaremenko, C.; Rodner, E.; Denzler, J.; Bohr, C.; Neumann, H.; Stelzle, F.; Maier, A. Automatic classification of cancerous tissue in laser endomicroscopy images of the oral cavity using deep learning. *Sci. Rep.* **2017**, *7*, 11979. [CrossRef] [PubMed]
28. Yasa, Y.; Çelik, Ö.; Bayrakdar, I.S.; Pekince, A.; Orhan, K.; Akarsu, S.; Atasoy, S.; Bilgir, E.; Odabaş, A.; Aslan, A.F. An artificial intelligence proposal to automatic teeth detection and numbering in dental bite-wing radiographs. *Acta Odontol. Scand.* **2021**, *79*, 275–281. [CrossRef] [PubMed]

applied sciences

MDPI

Review

Correlation between Temporomandibular Disorders and Tinnitus and Possible Treatment Strategies: Comprehensive Review

Angelo Michele Inchingolo [1,†], Alessio Danilo Inchingolo [1,†], Vito Settanni [1], Nicole De Leonardis [1], Merigrazia Campanelli [1], Grazia Garofoli [1], Stefania Benagiano [1], Giuseppina Malcangi [1], Elio Minetti [2], Andrea Palermo [3], Francesco Inchingolo [1,*], Gianna Dipalma [1,*,‡] and Assunta Patano [1,‡]

1 Department of Interdisciplinary Medicine, University of Bari "Aldo Moro", 70124 Bari, Italy; angeloinchingolo@gmail.com (A.M.I.); ad.inchingolo@libero.it (A.D.I.); v.settanni@libero.it (V.S.); nicoledeleonardis@outlook.it (N.D.L.); merigrazia.92@hotmail.it (M.C.); graziagarofoli.g@libero.it (G.G.); stefania.studiobenagiano@gmail.com (S.B.); giuseppinamalcangi@libero.it (G.M.); assuntapatano@gmail.com (A.P.)

2 Department of Biomedical, Surgical, Dental Science, University of Milan, 20161 Milan, Italy; elio.minetti@gmail.com

3 College of Medicine and Dentistry, Birmingham B4 6BN, UK; andrea.palermo2004@libero.it

* Correspondence: francesco.inchingolo@uniba.it (F.I.); giannadipalma@tiscali.it (G.D.); Tel.: +39-3312111104 (F.I.); +39-3396989939 (G.D.)

† These authors contributed equally to this work as first authors.

‡ These authors contributed equally to this work as last author.

Abstract: This study intends to investigate the relationship between otologic symptoms, in particular, tinnitus, and temporomandibular disorders (TMD). The literature studies during the previous 20 years had focused on the treatment and alleviation of the otologic symptoms and were limited to randomized clinical trials, case reports, and prospective studies. The following Boolean keywords, (tinnitus) AND (temporomandibular disorders OR temporomandibular therapy), were used in the databases of PubMed, Scopus, and Web of Science between 2003 and 9 May 2023 with an English language restriction. Results: The computerized search turned up 693 articles in total, and after eliminating duplicates, reviewing them, and determining their eligibility, 20 papers were included. Conclusion: The connections between temporomandibular TMD and tinnitus are numerous and intricate. It is unclear whether TMD could be the source of tinnitus or only its symptoms. Tinnitus may not always occur in persons with TMD, indicating that additional causes may potentially be involved in its occurrence. The precise mechanisms behind the link between TMD and tinnitus need to be clarified by additional study.

Keywords: tinnitus; craniomandibular disorders; craniomandibular disorders treatments; temporomandibular disorders; temporomandibular joint

Citation: Inchingolo, A.M.; Inchingolo, A.D.; Settanni, V.; De Leonardis, N.; Campanelli, M.; Garofoli, G.; Benagiano, S.; Malcangi, G.; Minetti, E.; Palermo, A.; et al. Correlation between Temporomandibular Disorders and Tinnitus and Possible Treatment Strategies: Comprehensive Review. *Appl. Sci.* **2023**, *13*, 8997. https://doi.org/10.3390/app13158997

Academic Editor: Rocco Franco

Received: 19 July 2023
Revised: 31 July 2023
Accepted: 4 August 2023
Published: 5 August 2023

1. Introduction

The joint connecting the temporal bone and the mandible is known as the temporomandibular joint (TMJ), and it consists of two components separated by an articular disc [1]. The TMJ plays a fundamental role in facilitating various jaw movements, including propulsion, lateral motion, and opening and closing of the mouth [1]. Temporomandibular disorders (TMD) represent a group of musculoskeletal conditions that affect the masticatory muscles, the temporomandibular joint, and related structures [1]. These disorders manifest through symptoms such as pain in the temporomandibular joint area or nearby, audible joint noises during mandibular movements, alterations in mandibular kinematics (such as deviations in opening, closing, laterality, and protrusion), and other related issues [1]. Tinnitus, a condition characterized by the perception of sound without an external source, and TMD are two complex medical issues that can interact in significant ways. Tinnitus, also known as "ringing in the ears", represents a subjective symptom, while temporomandibular disorders involve the complex musculoskeletal system of the jaw and surrounding structures [2].

In a study on the relationship between TMD otological symptoms, the data indicate that the prevalence of otologic symptoms (tinnitus, labyrinthitis, dizziness, loss of balance, and deafness) is 87% [3,4]. Tinnitus is the most common symptom and it is associated with 25–65% of TMD cases according to the scientific literature [5–11]. Tinnitus can sometimes be associated with somatic issues, even in the absence of hearing loss. This condition is known as "somatic tinnitus" (ST) or somatosensory tinnitus [7,12,13]. Among the common causes of ST are TMJ and head and neck diseases [7,12,14–18]. In cases of ST, the patient experiences altered somatosensory input from the temporomandibular joint or cervical spine [19]. Some studies have explored the connections between the cochlear nuclei (CN) and the somatosensory system in the cervical or temporomandibular region, providing a physiological basis for ST [20–22]. Afferent fibers from the dorsal root ganglia or trigeminal ganglion convey cervical somatosensory or temporomandibular information to the brain and central auditory system. ST may be suspected when at least one of the following is present in the patient's medical history: (a) history of head or neck trauma; (b) tinnitus associated with dental, jaw, or cervical spine manipulation; (c) recurrent pain episodes in the head, neck, or shoulder girdle; (d) temporal coincidence of the appearance or increase of both pain and tinnitus; (e) tinnitus worsening during inadequate postures while resting, walking, working, or sleeping; and (f) intense bruxism episodes during the day or night [23].

Simons et al. additionally suggested that tight myofascial trigger point bands in the deep masseter and lateral pterygoid muscles might be linked to middle ear injury [24].

Other authors claimed that the existence of this somatic modulation, as demonstrated by voluntary movements or particular resistance tests, is a key factor in the diagnosis of ST, if not the key factor [15,25]. The best way to identify ST is unclear because there are no established guidelines for clinical evaluation [19].

TMDs can cause various symptoms related to the ears and auditory system due to the anatomical proximity of the TMJ to these structures [26,27]. Sensations of closed ears and referred pain can be attributed to the malfunctioning of the Eustachian tube and tensor muscles [28,29]. Additionally, both the tensor muscles of the tympanic and palatine veil are innervated by the trigeminal; therefore, deep pain affecting the structures of this nerve can cause otological symptoms due to the brain stimulation [30–36].

It is unclear and still up for debate if craniomandibular problems and auditory symptoms relate to one another [37–39]. Some research indicates a decrease in otologic symptoms after receiving TMD treatments [8,37,40–46], while other studies do not support this association [47].

The aim of this study is to conduct a scoping review of the literature about the correlation between tinnitus and temporomandibular problems.

2. Materials and Methods

2.1. Protocol and Registration

This scoping review was conducted according to the standards of Preferred Reporting Items for Systematic Reviews and Meta-Analyses Extension for Scoping Reviews (PRISMA-ScR) [48].

2.2. Search Processing

A search on PubMed, Scopus, and Web of Science was performed (Table 1) to find papers that matched the topic of the relationship between tinnitus and TMD, dating from 1 January 2003 to 9 May 2023. The search strategy used the Boolean keywords: (tinnitus OR acuphen* OR acufen*) AND ("temporo mandibular disorder" OR "temporomandibular joint").

Table 1. Database search indicators.

Articles Screening Strategy	KEYWORDS: A: tinnitus; B: acuphen*; C: acufen*; D: temporo mandibular disorder; E: temporomandibular joint
	Boolean Indicators: (A OR B OR C) AND (D OR E)
	Timespan: from 1 January 2003 to 9 May 2023
	Electronic databases: PubMed; Scopus; WOS

2.3. Inclusion Criteria

The following inclusion criteria were considered: (1) studies that investigated the relationship between tinnitus and temporomandibular disorder, (2) randomized clinical trials, retrospective and observational studies, (3) English language, (4) full text available.

Papers that did not match the above criteria were excluded.

The review was conducted using the PCC criteria:

- Population: adults, male and female, with tinnitus and TMD,
- Concept: correlation between tinnitus and TMD,
- Context: public health care.

2.4. Exclusion Criteria

The exclusion criteria were as follows: (1) animal studies; (2) in vitro studies; (3) off-topic; (4) reviews, letters, or comments to editors; (5) no English language.

2.5. Data Processing

Five reviewers (N.D.L., G.G., V.S., M.C., and S.B.) independently consulted the databases to collect the studies and rated their quality based on selection criteria. The selected articles were downloaded into Zotero (version 6.0.15). Any divergence between the reviewers was settled by a discussion with a senior reviewer (F.I.).

3. Results

Study Selection and Characteristics

The electronic database search identified a total of 693 articles (Scopus N = 355, PubMed N = 191, Web of Science N = 147), and no articles were included through the hand search.

After the deletion of duplicates, 376 studies were screened by evaluating the title and abstract, focusing on the association between tinnitus and TMD.

There were 323 articles that did not meet the inclusion criteria (316 off topic, 33 review, four not in English) and two reports not retrieved, leading to 21 records being selected. After an eligibility check, one record was eliminated and 20 records were selected for qualitative analysis. The selection process and the summary of selected records were shown in Figure 1 and Table 2, respectively.

Among the included studies, eight papers examined the correlation between tinnitus and TMD, suggesting a causal role of TMDs in the onset and maintenance of tinnitus [1,14,15,49–53].

One study investigated the relationship between tinnitus and subtypes of TMD in patients with TMD. Patients with TMD might have a higher incidence of tinnitus when they present with PTF (type 1) [54].

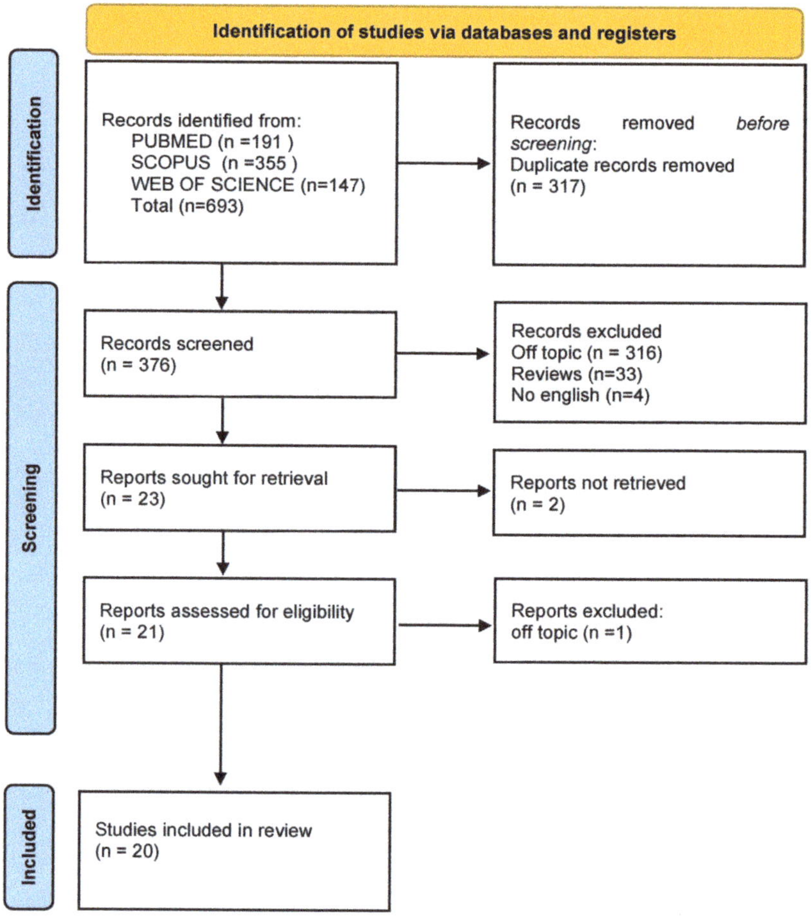

Figure 1. Literature search Preferred Reporting Items for Systematic Reviews and Meta-Analyses (PRISMA) flow diagram and database search indicators.

Another study, on the other hand, indicated a higher incidence of tinnitus among patients with TMD, potentially related to a reduction in glenoid fossa depth [55].

Four studies have evaluated the treatment effect of temporomandibular joint disorders when tinnitus is found, showing improvement after therapy [3,56–58].

Several authors have highlighted the psychological implications of TMDs and tinnitus, evaluating the incidence of stress on the onset of these pathologies and the impact on treatment. In addition, the incidence of depression in these patients has been discussed [50–53,59–62].

Two studies explored gender differences in the onset of TMD [52,61]. Female patients with TMD had a higher incidence of risk than male patients with TMD [52].

One paper described two cases of TMJ herniation that had tinnitus as an initial symptom and were treated differently [63].

Finally, a paper in 2020 described a rare case of tinnitus induced by chewing [64].

Table 2. Descriptive summary of item selection.

Authors	Type of Study	Study Design	Conclusion
Sobhy et al., 2004 [56]	A case-control study	Fifteen patients with dysfunctional myofascial pain syndrome and ear symptoms make up group (A). Fifteen patients in group (B) with disc dislocation, click, otalgia, or tinnitus. Occlusal splints, muscle relaxants, and anti-inflammatory medications were used to alleviate jerks when opening or closing the mouth as well as otalgia or tinnitus.	Before starting treatment, all of the patients in group (A) experienced otalgia. Ten patients were pain-free following therapy, five of whom saw a significant reduction in pain from severe to mild. Eight patients in group (B) no longer had pain, and seven saw a reduction in pain from severe to mild.
Webster et al., 2011 [57]	A prospective cohort study	The patients' age, gender, and tinnitus characteristics—the location of the symptom and the duration of time—were examined and an audiometric assessment was performed. Tinnitus intensity was assessed by a digital analogue scale before and after TMD treatment.	There was a significant decrease in tinnitus recognition by patients undergoing treatment for temporomandibular dysfunction.
Vielsmeier et al., 2011 [49]	A case-control study	In this study, the investigators compared 61 tinnitus patients without subjective complaints to a group of 30 patients with TMJ and tinnitus. With relation to demographic and clinical parameters, 61 patients with tinnitus but no subjective complaints were studied for mandibular joint dysfunction.	Patients with tinnitus who also had TMD had better hearing and tended to be female. In the sample that was looked at, 28% of the patients had TMD, while 72% did not.
Hilgenberg et al., 2012 [51]	An observational study	By assigning a score ranging from 0 to 100, the Mann–Whitney test was performed to examine the correlation between the severity of otological symptoms, the level of depression, and tinnitus.	In the tinnitus group, 40% of patients had severe depression, 30% had moderate depression, and 58% had typical levels of the illness. High tinnitus severity levels were substantially correlated with pain intensity.

Table 2. *Cont.*

Authors	Type of Study	Study Design	Conclusion
Vielsmeier et al., 2012 [14]	A retrospective study	The demographic, tinnitus, and audiological traits of 1204 tinnitus patients from the Tinnitus Research Initiative (TRI) Database with and without subjective TMJ symptoms were collected and compared using questionnaires and numerical ratings.	In tinnitus patients with TMD, traditional risk factors for tinnitus such as older age and male gender are less important. Tinnitus with TMD represents a subgroup of tinnitus that is most frequently masked by somatic maneuvers and music or sound stimulation
Akhter et al., 2013 [50]	A randomized clinic trial	Students with TMD symptoms were divided into seven groups: clicking only, TMJ pain only, difficulty opening the mouth only, clicking and pain, clicking and difficulty opening the mouth, difficulty opening the mouth and pain, and combination of three symptoms. The participants in the control group did not have any TMD symptoms.	TMD symptoms were shown to be substantially linked to auditory symptoms and headache.
Fernandes et al., 2013 [60]	A cross-sectional, clinically based study	The Temporomandibular Disorders Research Diagnostic Criteria TMD was classified using Axis I, while self-reported tinnitus and depression levels were scored using Axis II.	Painful temporomandibular problems, severe depression, and self-reported tinnitus are all strongly linked.
Ward et al., 2015 [15]	A prospective cohort study	A questionnaire with a series of questions was used to gather information about somatic tinnitus, and a score from 0 to 100 was assigned based on how severe it was.	In the questionnaire, 108 out of the 671 individuals claimed they could control their tinnitus using somatic exercises. Therefore, 16.1% of the population had tinnitus.
Lee et al., 2016 [52]	A retrospective population-based cohort study	Patients with TMJ disorder suffered from hearing loss, and degenerative and vascular ear disorders were compared with the control patients.	Higher age, hearing loss, and male gender have all been identified as risk factors for tinnitus. Female TMD patients had a greater risk incidence than male TMD patients.

Table 2. *Cont.*

Authors	Type of Study	Study Design	Conclusion
Çakur et al., 2016 [54]	A comparative study	CBCT pictures were examined based on PTF subtype diagnosis.	PTF (type 1) might be associated with an increased incidence of tinnitus in patients with TMD.
Kusdra et al., 2018 [1]	A retrospective study	A number of 485 medical records and data (gender, age, and otologic symptoms) were collected and analyzed.	A high prevalence of otological symptoms (in particular tinnitus) was found in TMD patients.
Lim, et al., 2019 [63]	A clinical case	The authors describe two cases of TMJ herniation through IF that presented with tinnitus and received different treatments.	In the case of surgery, the bone can be obliterated using tragal cartilage or the patient can be observed closely on an outpatient basis.
Edvall et al., 2019 [62]	An observational study	The online survey consisted of a combination of standardized questionnaires.	Stress and the patients' psychophysical state is an important risk factor in the increase of TMD—tinnitus.
Choi et al., 2020 [64]	A clinical case	Case of a 75-year-old patient who complained of chewing-induced popping tinnitus on the left side and echoendoscope examination revealed that the left tympanic membrane was protruding outwards due to due to the clenching of the teeth.	Mastitication-induced tinnitus is one of the probable signs of spontaneous ejection of TMJ tissue into the middle ear through a prolonged tympanic hole. Clenching the patient's teeth causes outward bulging of the TMJ, which is accompanied with clicking tinnitus.
Van der Wal et al., 2022 [3]	A randomized clinical trial	A total of 80 patients with TMD were enrolled from a tinnitus clinic. Patients were randomly assigned to either the physical therapy with occlusal splints group or the control group.	A treatment of TMD pain that allows for a reduction in tinnitus severity in patients with somatic tinnitus attributed to the TMJ.

Table 2. *Cont.*

Authors	Type of Study	Study Design	Conclusion
Lavinsky et al., 2020 [53]	A comparative study	A number of 53 adult patients with bilateral or unilateral TMD participated in the study, 30 of whom had tinnitus and 23 of whom did not.	The most frequent finding in both groups (24 individuals with tinnitus versus 15 without) was displacement of the articular disc.
Van der Wal et al., 2020 [61]	A retrospective study	Treatments used: HD-tDCS over 3 weeks' time in six sessions, psychological therapies were provided with a TRT +CBT combination and a maximum of 18 sessions of orofacial therapy in 9 weeks.	Males improved in the TRT +CBT group. Females improved in the HD-tDCS ($p = 0.0009$) and orofacial therapy group.
Plaza-Manzano et al., 2021 [58]	Randomized clinic trial	A secondary analysis of a clinical study was performed in 61 subjects with TMD-related tinnitus to determine the efficacy of incorporating cervico-mandibular manual treatment into exercise coupled with an education program.	In people with TMD-related tinnitus who received physical treatment, baseline tinnitus severity, and localized pressure pain threshold (PPT) over the temporalis muscle were predictive of clinical results.
Koparal et al., 2022 [55]	An observational study	TC was used to assess 82 patients with temporomandibular disorders (TMD) (40 patients without tinnitus and 42 patients with tinnitus).	The study revealed that decreased glenoid fossa depth may be associated with a higher incidence of tinnitus in TMD patients.
Kong et al., 2022 [59]	A cross-sectional study	Participants were divided into four groups: those with no TMD and no tinnitus, those with TMD but no tinnitus, those with TMD but no tinnitus, and those with both TMD and tinnitus.	TMD and tinnitus both have a significant influence on HRQoL in the Korean population.

4. Discussion

Some studies have investigated the correlation between TMD and tinnitus by evaluating whether treatment of TMD could benefit otological symptoms.

4.1. Association between TMD and Tinnitus

Vielsmeier et al. (2011) investigated the association between TMD and tinnitus by comparing patients with tinnitus and proven TMJ dysfunction to people with tinnitus but no subjective symptoms of TMJ dysfunction [49]. The results showed that tinnitus patients with TMJ disorder had better hearing function, were younger, experienced tinnitus at a younger age, and were more frequently female. Additionally, their perceived tinnitus loudness was lower, and more of them could modulate their tinnitus by moving their jaw or neck. This suggests that TMJ dysfunction might play a causal role in the generation and persistence of tinnitus. Therefore, the treatment of TMJ disorder could be a promising strategy for addressing tinnitus from a causal perspective [49].

In another piece of research conducted by the same authors, individuals with tinnitus and TMD were more likely to be female (with TMD: 54%; without: 33%), significantly younger, and with an earlier onset of tinnitus than those without TMD [49]. In addition, patients with tinnitus and TMD were more often able to mask tinnitus with sounds or music (with TMJ disorder: 85%; without: 74%) and could more often modulate tinnitus with somatic maneuvers (with TMJ disorder: 48%; without: 30%) [14].

Kusdra et al. (2018) investigated the prevalence of ear-related symptoms in patients with temporomandibular disorders (TMD) [1]. The researchers examined medical records of 485 patients at a specialized center and found that 87% of TMD cases reported otological symptoms, including tinnitus and ear fullness. The study provided evidence supporting the connection between TMD and otological symptoms, despite the symptoms not being directly caused by the ear [1]. A different study by Ward et al. investigated somatic tinnitus, which involves modulating tinnitus through somatic maneuvers, and showed a higher proportion of individuals with pulsatile tinnitus (not related to the heartbeat), under the age of 40, experiencing variations in tinnitus loudness, and having temporomandibular joint (TMJ) disorder [15].

Based on the data obtained from the studies examined, the prevalence of tinnitus among TMD patients is estimated to be 37.4% [15,49,53,55].

4.2. Factors Influencing TMD and Tinnitus Association

Some authors identified specific factors influencing the TMD and tinnitus association. Çakur et al. (2016) suggested that a certain type of TMD (PTF type 1) might be associated with an increased incidence of tinnitus. The authors employed a tac cone beam (CBCT) to assess the relationship between PTF subtype and tinnitus in TMD patients [54]. The study of Kijak et al. aimed to investigate the relationship between the displacement of the condyle in the temporomandibular joint (TMJ), the structure and position of the petrotympanic fissure (PTF), and comorbid tinnitus in patients with temporomandibular joint and muscle disorder (TMD) [65]. The researchers enrolled 331 TMD patients (268 women and 63 men), with an average age of approximately 40 years for women and 38 years for men. They used imaging studies of the facial part of the skull to analyze the TMJ and PTF configurations; around 10% of the TMD patients reported having tinnitus [65].

Lavinsky et al. (2020), using magnetic resonance imaging, found that patients with TMJ dysfunction and tinnitus presented a displacement of the joint disc with greater reduction than those with TMD but without tinnitus [53]. Furthermore, in the study of Jin Woo Choi et al. in 2020, the authors described a rare condition of chewing-induced tinnitus that occurs because of a defect in the normal ossification of the tympanic portion of the temporal bone that creates a communication pathway between the external auditory canal and the infratemporal fossa [64]. Instead, Koparal et al. aimed to use computed tomography (CT) to evaluate glenoid fossa depth and the horizontal angle of the ramus mandible in patients with TMD with tinnitus and without tinnitus to determine the association between TMD and tinnitus [55]. It was found that the glenoid fossa depth was decreased in patients with TMD and tinnitus compared to those with TMD only. This suggests a possible link between TMD and tinnitus due to the close anatomical relationship between the temporomandibular joint and the ear. Further research is needed to explore this association in more detail [55]. Another factor considered is the tympanic foramen or foramen of Huschke (FH), which is closed during the first five years of life [66]. The persistence of this defect can cause spontaneous herniation of the retrodiscal tissue of the TMJ in the middle ear with consequent mandibular clicking and tinnitus enlargement of the tympanic foramen, which can increase with age, resulting in tinnitus and hearing loss, as assessed by Lim et al. [63]. Assessment by high-resolution TBCT of the external auditory canal may be useful for diagnosis. Treatment in these cases requires surgery to correct the defect, which can be obliterated using tragus cartilage [67].

4.3. Study on Socio-Economic Factors and Psychological Variables

The study by Edvall et al. explored socio-economic factors, phenotypic characteristics, and psychological variables in tinnitus patients with or without TMJ complaints [62]. TMJ issues were more common in severe tinnitus cases, indicating a strong link. Subjects with TMJ complaints, mainly females, often attributed stress as the cause of their tinnitus. They experienced more severe tinnitus symptoms, poorer psychological well-being, and lower quality of life. This author uses TSCHQ (tinnitus sample case history questionnaire) [62]. They also reported difficulty tolerating sounds, headaches, vertigo, dizziness, and neck pain. These individuals often had pulsating or tonal tinnitus and somatic modulation, which worsened with loud sounds and stress. Stress was suggested to contribute to the co-occurrence and development of severe tinnitus, emphasizing dental care and stress management in comprehensive treatment [62]. In 2013, Fernandes et al., in their study, already demonstrated a strong link between tinnitus self-report, chronic severe TMD, and severe depression. The RDC/TMD questionnaire (research diagnostic criteria for temporomandibular disorders) was the test used [60]. These authors agree with Kong et al. (2022), who investigated the relationships between TMD, tinnitus, and quality of life in adults (nineteen years or older) using a nationally representative sample [59]. Participants were assigned to four groups: those with no TMD and no tinnitus, those with TMD but no tinnitus, those with TMD but no tinnitus, and those with TMD and tinnitus. The TMD and tinnitus group referred to the major number of concerns in the normal activity, pain/discomfort, and anxiety/depression aspects [59]. Clinicians should comprehend the idea that tinnitus is a complicated, multidimensional developmental process encompassing numerous physical, psychosocial, and environmental components to avoid using overly simplified techniques when diagnosing and treating tinnitus patients [59,60]. Another study by Akhter et al. (2013) aimed to explore the connections between aural symptoms, headache, and depression with temporomandibular disorder (TMD) symptoms through a survey involving university students [50]. Finally, they reported that clicking-only TMD symptoms were linked to otalgia and depression and suggested the importance of considering a functional assessment of the stomatognathic system in individuals experiencing unexplained aural symptoms and headache [50]. Hilgenberg et al. (2012) investigated the prevalence of temporomandibular disorders (TMD) and otologic symptoms in individuals with and without tinnitus while also examining the impact of depression levels comparing a tinnitus group with a control group [51]. The study revealed that 85% of tinnitus patients showed signs of TMD, compared to 55% in the control group, and found a positive association between tinnitus severity, higher depression levels, and TMD [51].

4.4. Tinnitus Treatment

In 2020, Van der Wal et al. discovered significant gender differences in the treatment results of numerous tinnitus care approaches [61]. Women in this study population have demonstrated a better response to treatment after receiving successive HDtDCS (high-definition transcranial direct current stimulation) sessions. HD-tDCS is a unique non-invasive brain stimulation technology based on the idea that mild intensity electric currents applied to locations of the scalp generate underlying cerebral activity [61]. These findings are consistent with those of Frank et al. (2012), who found greater benefits from frontal tDCS in women [68]. Chun-Feng Lee et al. (2016) conducted retrospective cohort research in 2016 on patients aged 20 years and older who had recently been diagnosed with TMJ dysfunction [52]. The authors suggested that TMD may increase the risk of tinnitus, but they found age, sex, and comorbidities of hearing loss, noise impacts on the inner ear, and degenerative and vascular ear illnesses as demographic characteristics and comorbidities that may be related to tinnitus [52]. The incidence of tinnitus was greater in the TMJ disease cohort than in the control patient, and female TMD patients have a higher risk incidence than male TMD patients [52]. Sobhy et al. (2004) conducted a case-control study aiming to assess the aural manifestations in patients with TMD and the effect of TMD treatment on tinnitus. Thirty patients were divided into two groups: Group (A) with myofascial pain dysfunction syndrome, and Group (B) with disc displacement with reduction. In conclusion, this study highlights the importance of evaluating aural manifestations in TMD patients. The results suggest that conservative treatment of TMD can effectively alleviate ear symptoms and improve the overall condition in these patients. In fact, the incidence of ear symptoms (otalgia and tinnitus) significantly decreased after therapy in both groups [56]. The same results were obtained by Webster et al. (2011), who found a significant decrease in tinnitus symptoms after treating temporomandibular dysfunction [57]. The researchers conducted a prospective cohort study involving individuals with TMD and tinnitus, and after dental treatment for TMD, there was a significant decrease in the intensity of tinnitus. Notably, tinnitus completely disappeared in four (26.6%) patients [57]. Van der Wal et al. (2020) conducted a randomized clinical trial, demonstrating that treating TMD pain reduced tinnitus severity in patients with somatic tinnitus attributed to the temporomandibular joint. The same results were obtained by Webster et al. (2011), who found a significant decrease in tinnitus recognition after treating temporomandibular dysfunction. The researchers conducted a prospective cohort study involving individuals with TMD and tinnitus, and after dental treatment for TMD, there was a significant decrease in the intensity of tinnitus. Notably, tinnitus completely disappeared in four (26.6%) patients [57]. In 2020, Plaza-Manzano et al. evaluated the impact of physical, clinical, psychological, and psychophysical factors on treatment results following the use of cervico-mandibular manual therapy, education, and exercise in patients with TMD and tinnitus [58]. The study highlighted the importance of baseline tinnitus severity and localized PPT over the temporalis muscle as significant factors in predicting treatment outcomes for individuals with TMD-related tinnitus undergoing physical therapy. Other predictors, such as sex and quality of life, also played a role but were less influential in the treatment outcomes [58].

5. Conclusions

In conclusion, there are numerous and complex relationships between TMD and tinnitus. Even while some writers have questioned whether an issue with the TMJ is the actual cause of tinnitus or just one of its symptoms, research shows that there is a strong correlation between the two illnesses. This study reveals that not all patients with TMD will develop tinnitus, indicating that other factors may also contribute. Tinnitus is known to be associated with advanced age, male gender, and hearing loss.

Numerous alternative processes have been used to explain the connection between tinnitus and TMD. Tinnitus can be caused by the trigeminal nerve, which innervates the TMJ, modulating the activity of the central auditory pathway. Tinnitus development may also be influenced by structural anomalies, such as a shallower glenoid fossa or flaws in the ossification of the temporal bone. Additionally, there is evidence that stress and bruxism may share a same ethology with tinnitus and TMD.

Physical therapy, manual therapy, exercise, education, and surgical intervention are some of the different treatment modalities for TMJ-related tinnitus. Studies have demonstrated that a multidisciplinary strategy that incorporates several medicines can enhance clinical results. With some encouraging results, repetitive transcranial magnetic stimulation (rTMS) has also been researched as a potential treatment option for tinnitus.

According to data in the literature, occlusal splints do not contribute to the improvement of tinnitus symptoms.

Tinnitus has an impact on patients' quality of life, so a multidisciplinary treatment approach is necessary. A thorough approach to diagnosis and treatment is essential because both illnesses have been linked to significant levels of depression. Tinnitus is a complex condition and treating it effectively and enhancing patients' wellbeing requires an understanding of how it interacts with TMD.

To create more specialized and individualized treatment modalities, additional study is required to clarify the precise mechanisms behind the association between TMD and tinnitus. Large-scale, national-level investigations that take the influence on quality of life into consideration are especially necessary. To effectively serve patients with TMD and tinnitus and ultimately improve their quality of life, healthcare providers must increase our understanding of this complex link.

Author Contributions: Conceptualization, A.M.I., G.M. and V.S.; methodology, E.M. and N.D.L.; software, M.C. and A.P. (Andrea Palermo); validation, G.G., S.B. and A.D.I.; formal analysis, E.M.; investigation, N.D.L., V.S., G.G. and G.D.; resources, F.I.; data curation, A.M.I. and A.P. (Assunta Patano); writing—original draft preparation, A.M.I., A.D.I. and A.P. (Assunta Patano); writing—review and editing, G.M.; visualization, G.G., S.B. and V.S.; supervision, N.D.L.; project administration, M.C., F.I., G.D. and A.P. (Assunta Patano); funding acquisition, A.P. (Andrea Palermo), F.I., G.D. and A.M.I. All authors have read and agreed to the published version of the manuscript.

Funding: This research received no external funding.

Institutional Review Board Statement: Not applicable.

Informed Consent Statement: Not applicable.

Data Availability Statement: Not applicable.

Conflicts of Interest: The authors declare no conflict of interest.

Abbreviations

CN	Cochlear nucleus
FH	Foramen of Huschke
MRI	Magnetic Resonance Imaging
TBCT	Tetrahedron Beam Computed Tomography
TC	Computed Tomography
TMD	Temporomandibular Disorders
TMJ	Temporomandibular Joint
PTF	Petrotympanic Fissure
HD tDCS	High-Definition Transcranial Direct Current Stimulation

References

1. Kusdra, P.M.; Stechman-Neto, J.; de Leão, B.L.C.; Martins, P.F.A.; de Lacerda, A.B.M.; Zeigelboim, B.S. Relationship between Otological Symptoms and TMD. *Int. Tinnitus J.* **2018**, *22*, 30–34. [CrossRef]
2. Baguley, D.; McFerran, D.; Hall, D. Tinnitus. *Lancet* **2013**, *382*, 1600–1607. [CrossRef] [PubMed]
3. van der Wal, A.; Michiels, S.; Van de Heyning, P.; Gilles, A.; Jacquemin, L.; Van Rompaey, V.; Braem, M.; Visscher, C.M.; Topsakal, V.; Truijen, S.; et al. Reduction of Somatic Tinnitus Severity Is Mediated by Improvement of Temporomandibular Disorders. *Otol. Neurotol.* **2022**, *43*, e309–e315. [CrossRef]
4. Kuttila, S.; Kuttila, M.; Bell, Y.L.; Alanen, P.; Jouko, S. Aural Symptoms and Signs of Temporomandibular Disorder in Association with Treatment Need and Visits to a Physician. *Laryngoscope* **1999**, *109*, 1669. [CrossRef] [PubMed]
5. Silveira, A.M.; Feltrin, P.P.; Zanetti, R.V.; Mautoni, M.C. Prevalence of Patients Harboring Temporomandibular Disorders in an Otorhinolaryngology Departament. *Braz. J. Otorhinolaryngol.* **2007**, *73*, 528–532. [CrossRef]
6. Mummolo, S.; Nota, A.; Marchetti, E.; Padricelli, G.; Marzo, G. The 3D Tele Motion Tracking for the Orthodontic Facial Analysis. *Biomed. Res. Int.* **2016**, *2016*, 4932136. [CrossRef]
7. Wu, C.; Stefanescu, R.A.; Martel, D.T.; Shore, S.E. Tinnitus: Maladaptive Auditory-Somatosensory Plasticity. *Hear Res.* **2016**, *334*, 20–29. [CrossRef]
8. Williamson, E.H. Interrelationship of Internal Derangements of the Temporomandibular Joint, Headache, Vertigo, and Tinnitus: A Survey of 25 Patients. *Cranio* **1990**, *8*, 301–306. [CrossRef]
9. Parker, W.S.; Chole, R.A. Tinnitus, Vertigo, and Temporomandibular Disorders. *Am. J. Orthod. Dentofacial. Orthop.* **1995**, *107*, 153–158. [CrossRef] [PubMed]
10. Vernon, J.; Griest, S.; Press, L. Attributes of Tinnitus That May Predict Temporomandibular Joint Dysfunction. *Cranio* **1992**, *10*, 282–287, discussion 287–288. [CrossRef]
11. Pasini, M.; Giuca, M.R.; Ligori, S.; Mummolo, S.; Fiasca, F.; Marzo, G.; Quinzi, V. Association between Anatomical Variations and Maxillary Canine Impaction: A Retrospective Study in Orthodontics. *Appl. Sci.* **2020**, *10*, 5638. [CrossRef]

12. Ralli, M.; Greco, A.; Turchetta, R.; Altissimi, G.; de Vincentiis, M.; Cianfrone, G. Somatosensory Tinnitus: Current Evidence and Future Perspectives. *J. Int. Med. Res.* **2017**, *45*, 933–947. [CrossRef]
13. Shore, S.; Zhou, J.; Koehler, S. Neural Mechanisms Underlying Somatic Tinnitus. *Prog. Brain Res.* **2007**, *166*, 107–123. [CrossRef] [PubMed]
14. Vielsmeier, V.; Strutz, J.; Kleinjung, T.; Schecklmann, M.; Kreuzer, P.M.; Landgrebe, M.; Langguth, B. Temporomandibular Joint Disorder Complaints in Tinnitus: Further Hints for a Putative Tinnitus Subtype. *PLoS ONE* **2012**, *7*, e38887. [CrossRef] [PubMed]
15. Ward, J.; Vella, C.; Hoare, D.J.; Hall, D.A. Subtyping Somatic Tinnitus: A Cross-Sectional UK Cohort Study of Demographic, Clinical and Audiological Characteristics. *PLoS ONE* **2015**, *10*, e0126254. [CrossRef]
16. Bhatt, J.; Ghavami, Y.; Lin, H.W.; Djalilian, H. Cervical Spine Dysfunctions in Patients with Chronic Subjective Tinnitus. *Otol. Neurotol.* **2015**, *36*, 1459–1460. [CrossRef] [PubMed]
17. Ralli, M.; Altissimi, G.; Turchetta, R.; Mazzei, F.; Salviati, M.; Cianfrone, F.; Orlando, M.P.; Testugini, V.; Cianfrone, G. Somatosensory Tinnitus: Correlation between Cranio-Cervico-Mandibular Disorder History and Somatic Modulation. *Audiol. Neurootol.* **2016**, *21*, 372–382. [CrossRef] [PubMed]
18. Levine, R.A.; Abel, M.; Cheng, H. CNS Somatosensory-Auditory Interactions Elicit or Modulate Tinnitus. *Exp. Brain Res.* **2003**, *153*, 643–648. [CrossRef]
19. Michiels, S.; Ganz Sanchez, T.; Oron, Y.; Gilles, A.; Haider, H.F.; Erlandsson, S.; Bechter, K.; Vielsmeier, V.; Biesinger, E.; Nam, E.-C.; et al. Diagnostic Criteria for Somatosensory Tinnitus: A Delphi Process and Face-to-Face Meeting to Establish Consensus. *Trends Hear.* **2018**, *22*, 2331216518796403. [CrossRef]
20. Lanting, C.P.; de Kleine, E.; Eppinga, R.N.; van Dijk, P. Neural Correlates of Human Somatosensory Integration in Tinnitus. *Hear. Res.* **2010**, *267*, 78–88. [CrossRef]
21. Shore, S.E. Plasticity of Somatosensory Inputs to the Cochlear Nucleus--Implications for Tinnitus. *Hear. Res.* **2011**, *281*, 38–46. [CrossRef] [PubMed]
22. Zhan, X.; Pongstaporn, T.; Ryugo, D.K. Projections of the Second Cervical Dorsal Root Ganglion to the Cochlear Nucleus in Rats. *J. Comp. Neurol.* **2006**, *496*, 335–348. [CrossRef] [PubMed]
23. Sanchez, T.G.; Rocha, C.B. Diagnosis and Management of Somatosensory Tinnitus: Review Article. *Clinics* **2011**, *66*, 1089–1094. [CrossRef] [PubMed]
24. Simons, D.G.; Travell, J.G.; Simons, L.S.; Travell, J.G. *Travell & Simons' Myofascial Pain and Dysfunction: The Trigger Point Manual*, 2nd ed.; Williams & Wilkins: Baltimore, MD, USA, 1999; ISBN 978-0-683-08363-7.
25. Biesinger, E.; Groth, A.; Höing, R.; Hölzl, M. Somatosensoric tinnitus. *HNO* **2015**, *63*, 266–271. [CrossRef] [PubMed]
26. Ciancaglini, R.; Loreti, P.; Radaelli, G. Ear, Nose, and Throat Symptoms in Patients with TMD: The Association of Symptoms According to Severity of Arthropathy. *J. Orofac. Pain* **1994**, *8*, 293–297.
27. Michelotti, A. Bell's Orofacial Pains: The Clinical Management of Orofacial Pain, 6th Edition (2005): Author: Jeffrey, P. Okeson Publisher: Quintessence Publishing Co. Ltd., New Malden, Surrey, UK Price: £47.00 ISBN: 0-86715-439-X. *Eur. J. Orthod.* **2005**, *27*, 532. [CrossRef]
28. Gargiulo Isacco, C.; Ballini, A.; Paduanelli, G.; Nguyen, K.C.D.; Schiffman, M.; Aityan, S.K.; Tapparo, O.; Khoa, N.; Dipalma, G.; Inchingolo, F.; et al. A Systematic Review on Persistent Trigeminal Artery, in Searching for a Therapeutic Solution to Idiopathic and Unresponsive Trigeminal Nerve Inflammations and Migraines. *J. Biol. Regul. Homeost. Agents* **2019**, *33*, 155–169.
29. Quinzi, V.; Scibetta, E.T.; Marchetti, E.; Mummolo, S.; Giannì, A.B.; Romano, M.; Beltramini, G.; Marzo, G. Analyze My Face. *J. Biol. Regul. Homeost. Agents* **2018**, *32*, 149–158.
30. Malkin, D.P. The Role of TMJ Dysfunction in the Etiology of Middle Ear Disease. *Int. J. Orthod.* **1987**, *25*, 20–21.
31. Sicher, H.; Du Brul, E.L. *Anatomia Oral de Sicher e DuBrul*, 8th ed.; Artes Medicas: São Paulo, Brazil, 1991.
32. Brookes, G.B.; Maw, A.R.; Coleman, M.J. 'Costen's Syndrome'—Correlation or Coincidence: A Review of 45 Patients with Temporomandibular Joint Dysfunction, Otalgia and Other Aural Symptoms. *Clin. Otolaryngol. Allied Sci.* **1980**, *5*, 23–36. [CrossRef]
33. Cox, K.W. Temporomandibular Disorder and New Aural Symptoms. *Arch. Otolaryngol. Head Neck Surg.* **2008**, *134*, 389–393. [CrossRef] [PubMed]
34. Ramírez, L.M.; Ballesteros, L.; Sandoval, G. Tensor Tympani Muscle: Strange Chewing Muscle. *Med. Oral Patol. Oral Cir. Bucal* **2007**, *12*, 96–100.
35. Penkner, K.; Köle, W.; Kainz, J.; Schied, G.; Lorenzoni, M. The Function of Tensor Veli Palatini Muscles in Patients with Aural Symptoms and Temporomandibular Disorder. An EMG Study. *J. Oral Rehabil.* **2000**, *27*, 344–348. [CrossRef]
36. Franz, B.; Anderson, C. The Potential Role of Joint Injury and Eustachian Tube Dysfunction in the Genesis of Secondary Ménière's Disease. *Int. Tinnitus J.* **2007**, *13*, 132–137. [PubMed]
37. Ren, Y.F.; Isberg, A. Tinnitus in Patients with Temporomandibular Joint Internal Derangement. *Cranio* **1995**, *13*, 75–80. [CrossRef] [PubMed]
38. Pekkan, G.; Aksoy, S.; HekImoglu, C.; Oghan, F. Comparative Audiometric Evaluation of Temporomandibular Disorder Patients with Otological Symptoms. *J. Cranio-Maxillofac. Surg.* **2010**, *38*, 231–234. [CrossRef]
39. Inchingolo, F.; Tatullo, M.; Marrelli, M.; Inchingolo, A.M.; Tarullo, A.; Inchingolo, A.D.; Dipalma, G.; Podo Brunetti, S.; Tarullo, A.; Cagiano, R. Combined Occlusal and Pharmacological Therapy in the Treatment of Temporo-Mandibular Disorders. *Eur. Rev. Med. Pharmacol. Sci.* **2011**, *15*, 1296–1300.

40. Bush, F.M. Tinnitus and Otalgia in Temporomandibular Disorders. *J. Prosthet. Dent.* **1987**, *58*, 495–498. [CrossRef]
41. Marasa, F.K.; Ham, B.D. Case Reports Involving the Treatment of Children with Chronic Otitis Media with Effusion via Craniomandibular Methods. *Cranio* **1988**, *6*, 256–270. [CrossRef]
42. Wright, E.F.; Bifano, S.L. Tinnitus Improvement through TMD Therapy. *J. Am. Dent. Assoc.* **1997**, *128*, 1424–1432. [CrossRef]
43. Wright, E.F.; Syms, C.A.; Bifano, S.L. Tinnitus, Dizziness, and Nonotologic Otalgia Improvement through Temporomandibular Disorder Therapy. *Mil. Med.* **2000**, *165*, 733–736. [CrossRef] [PubMed]
44. Gelb, H.; Gelb, M.L.; Wagner, M.L. The Relationship of Tinnitus to Craniocervical Mandibular Disorders. *Cranio* **1997**, *15*, 136–143. [CrossRef] [PubMed]
45. Steigerwald, D.P.; Verne, S.V.; Young, D. A Retrospective Evaluation of the Impact of Temporomandibular Joint Arthroscopy on the Symptoms of Headache, Neck Pain, Shoulder Pain, Dizziness, and Tinnitus. *Cranio* **1996**, *14*, 46–54. [CrossRef] [PubMed]
46. Björne, A. Assessment of Temporomandibular and Cervical Spine Disorders in Tinnitus Patients. *Prog. Brain Res.* **2007**, *166*, 215–219. [CrossRef] [PubMed]
47. Loughner, B.A.; Larkin, L.H.; Mahan, P.E. Discomalleolar and Anterior Malleolar Ligaments: Possible Causes of Middle Ear Damage during Temporomandibular Joint Surgery. *Oral Surg. Oral Med. Oral Pathol.* **1989**, *68*, 14–22. [CrossRef]
48. Tricco, A.C.; Lillie, E.; Zarin, W.; O'Brien, K.K.; Colquhoun, H.; Levac, D.; Moher, D.; Peters, M.D.J.; Horsley, T.; Weeks, L.; et al. PRISMA Extension for Scoping Reviews (PRISMA-ScR): Checklist and Explanation. *Ann. Intern. Med.* **2018**, *169*, 467–473. [CrossRef]
49. Vielsmeier, V.; Kleinjung, T.; Strutz, J.; Bürgers, R.; Kreuzer, P.M.; Langguth, B. Tinnitus with Temporomandibular Joint Disorders: A Specific Entity of Tinnitus Patients? *Otolaryngol. Head Neck Surg.* **2011**, *145*, 748–752. [CrossRef]
50. Akhter, R.; Morita, M.; Ekuni, D.; Hassan, N.M.M.; Furuta, M.; Yamanaka, R.; Matsuka, Y.; Wilson, D. Self-Reported Aural Symptoms, Headache and Temporomandibular Disorders in Japanese Young Adults. *BMC Musculoskelet Disord.* **2013**, *14*, 58. [CrossRef]
51. Hilgenberg, P.B.; Saldanha, A.D.D.; Cunha, C.O.; Rubo, J.H.; Conti, P.C.R. Temporomandibular Disorders, Otologic Symptoms and Depression Levels in Tinnitus Patients. *J. Oral Rehabil.* **2012**, *39*, 239–244. [CrossRef]
52. Lee, C.-F.; Lin, M.-C.; Lin, H.-T.; Lin, C.-L.; Wang, T.-C.; Kao, C.-H. Increased Risk of Tinnitus in Patients with Temporomandibular Disorder: A Retrospective Population-Based Cohort Study. *Eur. Arch. Otorhinolaryngol.* **2016**, *273*, 203–208. [CrossRef]
53. Lavinsky, D.; Lavinsky, J.; Setogutti, E.T.; Rehm, D.D.S.; Lavinsky, L. The Role of Magnetic Resonance Imaging of the Temporomandibular Joint to Investigate Tinnitus in Adults with Temporomandibular Joint Disorder: A Comparative Study. *Int. Arch. Otorhinolaryngol.* **2020**, *24*, e68–e72. [CrossRef] [PubMed]
54. Çakur, B.; Yaşa, Y. Correlation between Tinnitus and Petrotympanic Fissure Status among Patients with Temporomandibular Joint Dysfunction. *J. Oral Maxillofac. Surg.* **2016**, *74*, 47–52. [CrossRef]
55. Koparal, M.; Sirik, M.; Yavuz, G.Y.; Ege, B. Evaluation of the Relationship between Temporomandibular Joint Disorders and Tinnitus with Computed Tomography. *J. Stomatol. Oral Maxillofac. Surg.* **2022**, *123*, e199–e205. [CrossRef]
56. Sobhy, O.A.; Koutb, A.R.; Abdel-Baki, F.A.; Ali, T.M.; El Raffa, I.Z.; Khater, A.H. Evaluation of Aural Manifestations in Temporo-Mandibular Joint Dysfunction. *Clin. Otolaryngol. Allied Sci.* **2004**, *29*, 382–385. [CrossRef] [PubMed]
57. Evaluating the Effect of the Temporomandibular Disorder Treatment over Tinnitus. Available online: https://arquivosdeorl.org.br/additional/acervo_eng.asp?id=786 (accessed on 30 May 2023).
58. Plaza-Manzano, G.; Delgado-de-la-Serna, P.; Díaz-Arribas, M.J.; Rodrigues-de-Souza, D.P.; Fernández-de-Las-Peñas, C.; Alburquerque-Sendín, F. Influence of Clinical, Physical, Psychological, and Psychophysical Variables on Treatment Outcomes in Somatic Tinnitus Associated with Temporomandibular Pain: Evidence from a Randomized Clinical Trial. *Pain Pract.* **2021**, *21*, 8–17. [CrossRef] [PubMed]
59. Kong, G.-S.; Lee, S.-H.; Park, K.S.; Cho, J.-H.; Kim, K.-W.; Ha, I.-H. Association of Temporomandibular Disorders and Tinnitus with Health-Related Quality of Life: A Cross-Sectional Study Using the Fifth Korea National Health and Nutrition Examination Survey. *J. Oral Rehabil.* **2022**, *49*, 283–294. [CrossRef]
60. Fernandes, G.; Gonçalves, D.A.; Siqueira, J.T.T.D.; Camparis, C.M. Painful Temporomandibular Disorders, Self Reported Tinnitus, and Depression Are Highly Associated. *Arq. Neuropsiquiatr.* **2013**, *71*, 943–947. [CrossRef]
61. Van der Wal, A.; Luyten, T.; Cardon, E.; Jacquemin, L.; Vanderveken, O.M.; Topsakal, V.; Van de Heyning, P.; De Hertogh, W.; Van Looveren, N.; Van Rompaey, V.; et al. Sex Differences in the Response to Different Tinnitus Treatment. *Front. Neurosci.* **2020**, *14*, 422. [CrossRef]
62. Edvall, N.K.; Gunan, E.; Genitsaridi, E.; Lazar, A.; Mehraei, G.; Billing, M.; Tullberg, M.; Bulla, J.; Whitton, J.; Canlon, B.; et al. Impact of Temporomandibular Joint Complaints on Tinnitus-Related Distress. *Front. Neurosci.* **2019**, *13*, 879. [CrossRef]
63. Lim, K.H.; Jung, J.Y.; Rhee, J.; Choi, J. Temporomandibular Joint Herniation through the Foramen of Huschke with Clicking Tinnitus. *Eur. Ann. Otorhinolaryngol. Head Neck Dis.* **2019**, *136*, 497–499. [CrossRef]
64. Choi, J.W.; Nahm, H.; Shin, J.E.; Kim, C.-H. Temporomandibular Joint Herniation into the Middle Ear: A Rare Cause of Mastication-Induced Tinnitus. *Radiol. Case Rep.* **2020**, *15*, 125–127. [CrossRef] [PubMed]
65. Kijak, E.; Szczepek, A.J.; Margielewicz, J. Association between Anatomical Features of Petrotympanic Fissure and Tinnitus in Patients with Temporomandibular Joint Disorder Using CBCT Imaging: An Exploratory Study. *Pain Res. Manag.* **2020**, *2020*, e1202751. [CrossRef]

66. Mao, J.J.; Nah, H.-D. Growth and Development: Hereditary and Mechanical Modulations. *Am. J. Orthod. Dentofacial. Orthop.* **2004**, *125*, 676–689. [CrossRef] [PubMed]
67. Heffez, L.; Anderson, D.; Mafee, M. Developmental Defects of the Tympanic Plate: Case Reports and Review of the Literature. *J. Oral Maxillofac. Surg.* **1989**, *47*, 1336–1340. [CrossRef] [PubMed]
68. Frank, E.; Schecklmann, M.; Landgrebe, M.; Burger, J.; Kreuzer, P.; Poeppl, T.B.; Kleinjung, T.; Hajak, G.; Langguth, B. Treatment of Chronic Tinnitus with Repeated Sessions of Prefrontal Transcranial Direct Current Stimulation: Outcomes from an Open-Label Pilot Study. *J. Neurol.* **2012**, *259*, 327–333. [CrossRef]

applied sciences

MDPI

Review

Investigation on the Application of Artificial Intelligence in Prosthodontics

Abdulkhaliq Ali F. Alshadidi [1], Ahid Amer Alshahrani [1], Lujain Ibrahim N. Aldosari [2], Saurabh Chaturvedi [2,*], Ravinder S. Saini [1], Saeed Awod Bin Hassan [3], Marco Cicciù [4,*] and Giuseppe Minervini [5,*]

1 Dental Technology Department, COAMS, King Khalid University, Abha 62529, Saudi Arabia
2 Department of Prosthetic Dentistry, College of Dentistry, King Khalid University, Abha 62529, Saudi Arabia
3 Department of Restorative Dental Sciences, College of Dentistry, King Khalid University, Abha 62529, Saudi Arabia
4 Department of Biomedical and Dental Sciences and Morphofunctional Imaging, School of Dentistry, University of Messina, 98125 Messina, Italy
5 Multidisciplinary Department of Medical-Surgical and Odontostomatological Specialties, University of Campania "Luigi Vanvitelli", 80138 Naples, Italy
* Correspondence: survedi@kku.edu.sa (S.C.); mcicciu@unime.it (M.C.); giuseppe.minervini@unicampania.it (G.M.)

Abstract: Artificial intelligence (AI) is a contemporary, information-driven innovative technology. Prosthetic dentistry, also known as prosthodontics, is the restoration and reconstruction of missing teeth utilizing implants for permanent and removable prostheses. It enhances healthy soft and hard tissues, promoting oral health. This study examined the use of artificial intelligence in prosthodontics to diagnose abnormalities and create patient-specific prostheses. Two researchers searched Google Scholar, Scopus, PubMed/MEDLINE, EBSCO host, Science Direct, and Web of Science (MEDLINE, WOS, and KJD). Articles on AI in English were reviewed. We also collected the following broad article aspects: research and control groups, assessment methodology, outcomes, and quality rankings. This methodological study examined AI use in prosthodontics using the latest scientific findings. The findings were statistically evaluated using ANOVA. Titles and abstracts revealed 172 AI-related dentistry studies, which were analyzed in this research. Thirty-eight papers were eliminated. According to the evaluation, AI was found to have significantly increased in prosthodontics. Despite the vast number of studies documenting AI applications, the description of the data illustrated the latest breakthroughs in AI in prosthodontics, highlighting its use in automatically produced diagnostics, predicting analytics, and classification or verification tools.

Keywords: artificial intelligence; prosthodontics; ANOVA variance

Citation: Alshadidi, A.A.F.; Alshahrani, A.A.; Aldosari, L.I.N.; Chaturvedi, S.; Saini, R.S.; Hassan, S.A.B.; Cicciù, M.; Minervini, G. Investigation on the Application of Artificial Intelligence in Prosthodontics. *Appl. Sci.* **2023**, *13*, 5004. https://doi.org/10.3390/app13085004

Academic Editor: Andrea Scribante

Received: 13 February 2023
Revised: 14 March 2023
Accepted: 30 March 2023
Published: 16 April 2023

1. Introduction

Prosthodontics is a branch of dentistry that is considered both an art and a science. It is the art and science of diagnosing, planning, rehabilitating, and preserving the function, comfort, appearance, and health of the oral structures of patients with clinical problems caused by missing or deficient teeth and oral and maxillofacial tissues. It fulfils this purpose mostly by replacing missing teeth and associated structures with artificial ones [1–7]. Prosthodontics focuses on the treatment and construction of removable and fixed dental prostheses, as well as the preparation of finishing margins along the tooth for better extension and fitting of the crowns, implant surgery, and the construction of a maxillofacial prosthesis. It is also used to develop and maintain the relationship between the upper and lower jaws for a stable prosthesis [8–10]. AI can be very helpful in many different kinds of therapy.

Prosthodontics, or the study of dental prostheses, is an important subject that has a broad influence on different phases of a dentist's life. Advancements in digital dentistry positively influenced prosthodontics and led to exponential development in the field of materials used, diagnosis and treatment planning, and even in prosthesis fabrication. Several factors can contribute to the fabrication of dental prostheses.

First, it was suggested that the deterioration in dental morphology poses a threat to a person's physical, mental, and behavioral health [1,2]. Due to difficulty in biting (or breaking down food), the patient starts shifting the food choices to a softer one or even try to avoid food.

Second, tooth loss makes people more isolated because of concerns regarding social approval [3]. Arguably, having beautiful teeth helps you socialize more since people generally want their entire body to look good. Finally, having a tooth in the oral cavity helps to keep the tongue, lips, and cheeks in the right positions, giving the facial features a proper shape [4]. In prosthodontics, the replacement of missing teeth is possible with the help of implants and crowns [5–10].

Advancements in digital dentistry positively influenced prosthodontics and led to exponential development in the field of materials used, diagnosis and treatment planning, and even in prosthesis fabrication. Several factors can contribute to the fabrication of dental prostheses. When there is a loss of teeth, the construction of prostheses such as removable partial dentures, fixed dental prosthesis or implants provides an alternative [11–13]. The type of implant utilized, and the health of the remaining residual alveolar ridge, determine how rehabilitation can be carried out. Multiple techniques are required during this rehabilitative procedure [14,15].

Artificial intelligence (AI) applications are widely used in digitized daily life [16–19], such as in the form of online companions such as "Alexa" or "Siri", and they are used in many technical disciplines. AI algorithms are widely used in the medical field to analyze images by extracting features and performing target studies [20,21]. Over the past five years, there was an increase in studies on the usefulness and use of AI in radiology to enhance workflow and alleviate some of the radiologist's workload. Although dental technology is improving, it is still behind medical technology [22–31]. In conventional treatment, standardized digital dental methods are used. The desktop design and CAD/CAM fabrication became standard in healthcare and labs. AI application is a new concept that is starting to emerge as dental digitization advances [32]. Dentures for edentulous patients are difficult to create because of the high aesthetic and functional criteria that must be satisfied. The CAD/CAM software's machine learning can realign the teeth to restore the inter-maxillary connections. Artificial intelligence (AI) might help with precise color matching in difficult aesthetic circumstances involving a single central incisor or several front teeth. In implant prosthodontics, implant locations may be identified with the use of intraoral detectors, and this information can then be inputted into the CAD program in real-time. Artificial intelligence (AI) has the capacity to improve dental implant design and fabrication [32].

AI (artificial intelligence) applications are used in many parts of our digital lives [16–19], such as online assistants like "Alexa" and "Siri", and in many technical fields. In the medical field, AI algorithms are often used to analyze images by pulling out features and doing target studies [20,21]. In the past five years, there have been more studies on how AI can be used in radiology to improve workflow and make the radiologist's job easier. Dental technology is getting better, but it is still not as good as medical technology [22–31]. Standardized digital dental methods are used in traditional dental care.

In labs, CAD/CAM fabrication is now the norm. AI application is a new idea that is starting to take shape as digital dentistry gets better [32]. Dentures for people with no teeth are technique sensitive because and they have to meet strict aesthetic and functional standards. Machine learning in the CAD/CAM software can realign the teeth to fabricate complete dentures with accuracy. Artificial intelligence (AI) could help match colours exactly in case of missing central incisors or adjust the colour with remaining natural teeth. In implant prosthodontics, an intraoral detector can be used to find the location of implants. This information can then be put into a CAD program in real-time. AI has the potential to make it easier to design and make better dental implant prosthetics [32].

The present review is novel and very specific to the field of dentistry, specifically prosthetic dentistry. It fills the gap in the literature in a way that there are various reviews pertaining to the use of AI in dentistry, endodontics, and oral medicine, but there is a need to provide a review in prosthetic dentistry which is lacking in the literature. This lacuna is filled by this review, as it explains the use and enhances the value of AI in field of dentistry and, more specifically, to prosthetic dentistry. This review will help in explaining the use and working of AI in different fields of prosthodontics, which would ultimately help the patients and improve satisfaction with the prosthesis.

Humans are increasingly demanding, and always ready for more detailed and thorough medical/dental care; therefore, relying on artificial intelligence can improve the standard of care for patients. AI provides cutting-edge decision-supporting technologies in the discipline of prosthodontics. Therefore, this review aimed to explore and identify the various filed in prosthodontics where AI is used and how it is useful in providing prosthodontic care.

2. Methodology

With the advancement in the field of technology, AI plays an important role not only in our day-to-day life but also in the field of dentistry, as well as in enhancing prosthetic rehabilitation. The present review was conducted with the same purpose of exploring and identifying AI use in prosthodontics. This review was filed and registered in the International Database to Register Systematic Reviews (INPLASY) dated 24 December 2022. The DOI registration no. is 10.37766/inplasy 2022.12.0096. It followed the PRISMA 2020 [33,34] and Cochrane criteria [35]. Figure 1 represents the flow chart depicting the methodology used in this review.

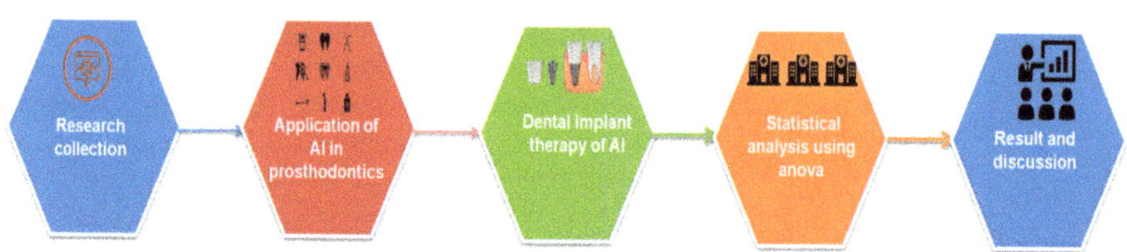

Figure 1. The flow of review methodology.

This systematic review followed PRISMA guidelines [33]. For the PICO question, the following statements were formulated:

1. What kinds of AI methods are currently being employed in the prosthodontic field?
2. How does AI enhance clinical decision-making and outcomes in prosthodontics?
3. What are the current clinical uses of AI in prosthodontics, and how well does it work as a diagnostic tool during the determination of the type of prosthesis required for a particular patient?

P—population: images of patients' and simulators' faces in two and three dimensions, as well as a variety of radiographs used during the prosthodontic rehabilitation procedure (including periapical, OPG, and cone-beam computed tomography). I—intervention: machine learning, natural language processing, and robots are used for the diagnosis, management, and prognosis assessment of prosthodontic procedures. C—comparison: image analysis, testing models, and automation algorithms involved in prosthodontic steps while using the digitalized method of prosthetic rehabilitation. O—outcome: several areas of prosthetic dentistry may benefit from the use of artificial intelligence, including performance, accuracy, precision, sensitivity, and clinical decision support (CDS).

Study design: English-language publications of both observational research (such as case–control and cohort studies) and experimental research (such as randomized controlled trials) were included in this analysis, which was basically concerned with the prosthodontic workflow.

2.1. Data Collection

Google Scholar, Scopus, PubMed/MEDLINE, Science Direct, EBSCO host, and Web of Science (MEDLINE, WOS, and KJD) provided scientific studies on AI in prosthodontics (WOS, KJD, MEDLINE, RSCI). Only prosthodontic papers were extensively collected. Each category was a mix of MeSH keywords and free-text words (Prosthodontics [Mesh]). The following parameters for inclusion and exclusion were followed: English-language articles, all evidence-based studies except expert opinions, and content from the previous four years (from 2017–18 to 30 September 2021–22). The following descriptions of studies were excluded: case reports with less than ten patients, editorials, reviews with no access to or full text of the document, and research articles on animal models.

Data Extraction

To choose studies and report AI in prosthetic dentistry, several publications were reviewed. The judgment of two review writers (S.C. and R.S.S., both prosthodontists) independently was considered to select and finalize the publications. Two steps were used to do this: first, matching was carried out for the titles and abstracts of all studies according to the inclusion criteria; next, the full texts of the publications that were deemed good from the first round of screening were examined. Information extraction: the two reviewers (L.A. and F.H.) performed their data extraction work independently at first, and then, compared and revised their results. The following information was gathered from different sources: details of the author(s), publication year and nationality, aim of the study, patient records, application of AI, test datasets, and outcome.

2.2. Study Selection

Based on the primary search keywords and phrases, 172 articles were identified. A total of 35 duplicates were eliminated, and 90 papers were screened based on their titles and abstracts, of which 21 were excluded. So, 69 reports were sought for retrieval. Eight items that could not be retrieved from the search were excluded. The eligibility of the remaining 61 studies was evaluated. A total of 32 reports were also excluded due to various reasons as described in the flow chart presented in Figure 2. Finally, 24 articles were included and examined. Figure 2 shows the PRISMA flow diagram of the literature review.

2.3. Study Characteristics

The studies included in the review were mainly systematic reviews, retrospective studies, and meta-analyses. (Table 1). Most of the research (66%) was published in the previous four years.

Table 1. Description of the study characteristic.

Study Title, Author Name and Publishing Year [Citation]	Country	Study Design	Journal	Aim of the Study
Artificial intelligence for fast and accurate 3-dimensional tooth segmentation on cone-beam computed tomography (EzEldeen et al. [36])	Belgium	CBCT	Journal of Endodontic	The purpose of this research was to create and verify an AI-driven instrument for automatic teeth segmentation using cone beam computed tomography (CBCT).
Machine learning and intelligent diagnostics in dental and orofacial pain management (Farook et al. [37])	Malaysia Saudi Arabia	Systematic review	Pain Research and Management	Machine learning was investigated to determine its clinical impact, efficacy, limitations, and results when compared to human diagnostics for identifying the root causes of dental and orofacial pain, and (6) bone and temporomandibular joint.

Table 1. *Cont.*

Study Title, Author Name and Publishing Year [Citation]	Country	Study Design	Journal	Aim of the Study
Uses of Different Machine Learning Algorithms for Diagnosis of Dental Carie (Talpur et al. [38])	Pakistan	DDC	Journal of Healthcare Engineering	The primary goal of this research was to systematically examine existing literature on the topic of how machine learning can affect dental caries. The PICO criteria will be used as a framework for this study's design.
Outcome measurements and quality of randomized controlled clinical trials of tooth-supported fixed dental prostheses (Limones et al. [39])	Spain	Systematic review	The Journal of Prosthetic Dentistry	This systematic review aimed to identify all primary and secondary outcome metrics in tooth-supported FDP RCTs. Secondary aims were to assess methodological quality using the Cochrane Collaboration's risk of bias instrument (RoB, v2.0) and reporting quality using a 16-item CONSORT evaluation tool through published reports.
Robot technology in dentistry (Van Riet et al. [40])	Netherlands	Systematic review	Dental Materials	The goal of this evidence-based review was to give dentists and researchers a bird's-seeing view of the features of the literature surrounding dental robotics projects.
A risk of bias tool and guideline to support reporting of pre-clinical dental materials research and assessment of systematic reviews (Delgado et al. [41])	Spain, Brazil, etc.	Systematic review	Journal of Dentistry	The purpose of this study was to aid in the reporting of future investigations and to enhance the assessment in systematic reviews.
Artificial Intelligence and Surgical Education: A Systematic Scoping Review of Interventions (Kirubarajan et al. [42])	Canada	Systematic review	Journal of Surgical Education	The purpose of this literature review was to consolidate the available research on the application of AI to surgical education.
Accuracy of intraoral scanners versus traditional impressions: A rapid umbrella review (Afrashtehfar et al. [43])	Switzerland	RU review	Journal of Evidence-Based Dental Practice	The primary purpose of this research was to (1) evaluate the reporting quality of titles and abstracts of the collected literature and (2) assess the truthfulness and precision of intraoral scanning (IOS) in dentistry based on recent secondary sources.
Accuracy of Digital Dental Implants Impression Taking with Intraoral Scanners Compared with Conventional Impression Techniques: A Systematic Review of In Vitro Studies (Albanchez-González [44])	Spain	Systematic Review	International Journal of Environmental Research and Public Health	This systematic study aimed to assess the in vitro accuracy of dental implant impressions produced with an intraoral scanner to those taken with more traditional methods.
Accuracy of removable partial denture metal frameworks fabricated by computer-aided design/computer-aided manufacturing method: A systematic review and meta-analysis (MAI et al. [45])	Republic of Korea	Systematic Review	Journal of Evidence Based Dental Practice	The purpose of this systematic review and meta-analysis was to evaluate the precision of removable partial denture (RPD) frameworks made with CAD/CAM systems to those made with traditional casting techniques.
Communication tools and patient satisfaction: A scoping review (Touati et al. [46])	Switzerland	Scoping review	Journal of Esthetic and Restorative Dentistry	This exploratory research evaluated aesthetic dentistry communication. Various communication tools can incorporate patients in SDM. Few know how dental communication technology improves patient satisfaction. Medline, Embase, Cochrane, and World Science were searched for patient satisfaction research.

Table 1. *Cont.*

Study Title, Author Name and Publishing Year [Citation]	Country	Study Design	Journal	Aim of the Study
Robotic applications in orthodontics: Changing the face of contemporary clinical care (Adel et al. [47])	Egypt and India	CFCC	BioMed research international	This review covered eight orthodontic domains: (1) robotic dental assistants; (2) robotics in orthodontic diagnosis and simulation; (3) robotics in patient education, teaching, and training; (4) wire bending and customized appliance robotics; (5) nanorobots/microrobots for tooth movement acceleration and remote monitoring; (6) robotics in maxillofacial surgeries and implant placement; (7) automated aligner production robotics; and (8) TMD rehabilitative robotics.
A review of 3D printing in dentistry: Technologies, affecting factors, and applications (Tian et al. [48])	China and Republic of Korea	Technological application	Scanning	Three-dimensional printing has uses in dentistry, including prosthodontics, oral and maxillofacial surgery, and oral implantology. The 3D printing review is practical and scientific.
Economic Evaluations of Preventive Interventions for Dental Caries and Periodontitis: A Systematic Review (Nguyen et al. [49])	Australia	Systematic Review	Applied Health Economics and Health Policy	For the purpose of providing a critical analysis of the techniques employed in comprehensive economic assessments of preventative therapies for dental caries and periodontitis.
Rehabilitation with dental prostheses and its influence on brain activity: A systematic review (Costa et al. [50])	Brazil	Systematic Review	The Journal of Prosthetic Dentistry	The aim was to determine whether or not oral prosthesis rehabilitation was associated with an increase in regional brain activity.
Detection of Caries under Fixed Prosthodontic Restorations Using Cone-beam CT: A Meta-analysis (Sivaramakrishnan et al. [51])	Bahrain	Meta-analysis	International Journal of Prosthodontics and Restorative Dentistry	This literature review sought to synthesize the data on cone-beam computed tomography's (CBCT) ability to identify caries under fixed restorations.
Intraoral scanning devices applied in fixed prosthodontics (Abad-Coronel et al. [52])	Cuenca	Dentistry	Acta Sci Dent Sci	Intraoral scanning devices applied in fixed prosthodontics (Abad-Coronel et al. [52]).
Efficacy of deep convolutional neural network algorithm for the identification and classification of dental implant systems, using panoramic and periapical radiographs: A pilot study (Lee, 2020 [53])	Republic of Korea	Retrospective Cohort Study	Medicine	The aim of the current study was to evaluate the efficacy of deep CNN algorithm for the identification and classification of dental implant systems.
Artificial intelligence in fixed implant prosthodontics: A retrospective study of 106 implant-supported monolithic zirconia crowns inserted in the posterior jaws of 90 patients (Lerner et al., 2020 [54])	Germany	Retrospective Cohort Study	BMC Oral Health	Purpose of this retrospective clinical study is to present a protocol for the use of AI to fabricate implant-supported monolithic zirconia crowns cemented on customized hybrid abutments, via a full digital workflow.
Predicting the debonding of CAD/CAM composite resin crowns with AI (Yamaguchi et al., 2019 [55])	Japan	Retrospective Cohort Study	J. Dent. Res.	The aim of this study was to assess the validity of deep learning with a CNN method to predict the debonding probability of CAD/CAM composite resins restorations from 2D images captured from 3D STL models of a die scanned by a 3D oral scanner.
Detection and diagnosis of dental caries using a deep learning-based convolutional neural network algorithm (Lee et al., 2018 [56])	Republic of Korea	Retrospective Cohort Study	J. Dent. Res.	The aim of the study was to evaluate the efficacy of deep CNN algorithms for detection and diagnosis of dental caries on periapical radiographs.

Table 1. *Cont.*

Study Title, Author Name and Publishing Year [Citation]	Country	Study Design	Journal	Aim of the Study
Diagnosis and prediction of periodontally compromised teeth using a deep learning-based convolutional neural network algorithm (Lee et al., 2018 [57])	Republic of Korea	Retrospective Cohort Study	Journal of Periodontal Implant	The aim of the current study was to develop a computer-assisted detection system based on a deep CNN algorithm and to evaluate the potential usefulness and accuracy of this system for the diagnosis and prediction of periodontally compromised teeth.
Artificial Neural Networks as a powerful numerical tool to classify specific features of a tooth based on 3D scan data (Raith et al., 2017 [58])	Germany	Retrospective Cohort Study	Comput. Bio. Med.	The hypothesis was that tooth classification algorithms based on ANNs are capable of classifying teeth with sufficient accuracy for potential use in clinical practice in order to improve digital workflow in dental prosthetics.
Evaluation of a Novel Computer Color Matching System Based on the Improved BackPropagation Neural Network Model. J. P (Wei et al., 2018 [59])	China	Retrospective Cohort Study	J. Prosthodont.	[The aim of this study was] to explore the feasibility of a novel computer color-matching system based on the improved back-propagation neural network model by comparing it with the traditional method.

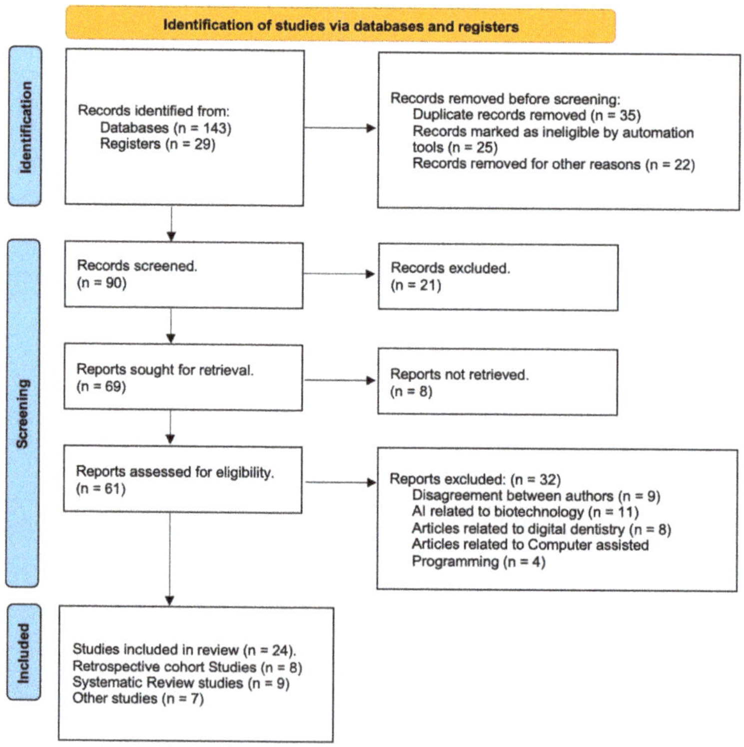

Figure 2. Flowchart of the Prisma.

Assessment of the Risk of Bias in Included Studies

The papers that were included had their bias risk evaluated separately by the two researchers. The research was considered to have a low risk of bias if it provided thorough information on 80% or more of the relevant parameters. There was a moderate probability of bias in studies if the supplied information corresponded to less than 50–70% of the relevant parameters and a high risk of bias if more than 50% of the relevant parameters were left unanswered [60,61].

2.4. Statistical Analysis

Prosthodontics AI research evaluation employed ANOVA [62]. This demonstrates the use of the AI model and its impact on how prosthodontics is impacted, sustained, and upgraded. The standard A-statistic for ANOVA was determined by dividing the sum of the squared averages of the void sample with the anthropocentric principle by the entire model. Each variance was considered equally when determining the parameters using the approach of reducing squared errors.

3. Result

3.1. Descriptive Analysis

Eight publications showed various applications of AI within the discipline of prosthodontics, despite their varied origins. Eight studies investigated the development and use of various AI systems [63–71]. Two studies investigated the use of a particular CAD program's built-in AI system for creating prostheses [66]. This review employed a descriptive method to analyze the data supplied by the included research. Table 1 provides a synopsis of the data extraction procedure. Each one of the included studies used a non-randomized, retrospective cohort design. Four of these publications focused on CBCT [64,68,69]. In total, there were four studies that examined neural network models, one that employed a forecasting model, and one that made use of the built-in AI and algorithms of a widely used analytical CAD program. Prosthodontics is a discipline where AI is steadily making inroads and will keep doing so. Development was made in the field, with a focus on CBCT (cone beam computed tomography) and 3D scans for more accurate implant prosthodontic diagnosis. Excellent outcomes were achieved by using AI to precisely create surgical templates and assess bone integrity. This section examines how AI may affect prosthodontics.

3.2. Statistical Analysis by ANOVA

The impact of AI deployment on prosthodontics was examined using ANOVA. The ANOVA test results for the AI application in prosthodontics are shown in Table 2.

Table 2. ANOVA test of the AI application on prosthodontics.

Features	*p*-Value
Diagnostic accuracy	0.022
Error reduction	0.027

All data that could be obtained from the articles were tallied. All papers that met the inclusion criteria were evaluated for potential bias using the quality of evidence scale (Table 3).

Table 3. Study bias assessment. (* shows the presence of the factor checked.)

Stars	Lerner, H.; Mouhyi, J.; Admakin, O.; Mangano, F. [54]	Lee, J.-H.; Jeong, S.-N. [53]	Lee, J.-H.; Kim, D.-H.; Jeong, S.-N.; Choi, S.-H. [56]	Yamaguchi, S.; Lee, C.; Karaer, O.; Ban, S.; Mine, A.; Imazato, S. [55]	Raith, S.; Vogel, E.P.; Anees, N.; Keul, C.; Güth, J.-F.; Edelhoff, D.; Fischer, H. [58]	Lee, J.-H.; Kim, D.-h.; Jeong, S.-N.; Choi, S.-H. [57]	Wei, J.; Peng, M.; Li, Q.; Wang, Y. [59]
Determination (Max. 4 Stars)	***	**	**	***	**	**	*
Comparison (Max. 2 Stars)	-	-	-	*	-	-	-
Outcome (Max. 4 Stars)	*	*	*	*	*	*	*

3.3. Diagnostic Accuracy

All subsequent therapies and suggestions were based on the results of the diagnostic accuracy examination. A thorough evaluation of the patient's teeth, musculature, and associated tissues are required. Challenges with missing teeth, complications with replacement, and arch diagnosis for implant surgery are necessary. AI significantly increases diagnosis accuracy.

The improvement in diagnostic accuracy explains the improvement in diagnosis during prosthodontic rehabilitation and showed that the use of AI was directly proportional to accuracy. Here, diagnosis includes the determination of the problem, the decision on the type of prosthesis, and finalizing the design and related components in removable and fixed prostheses.

AI and machine learning emerged as powerful tools for assisting diagnosis, determination of the type of prosthesis requirement, the development and positioning of clasps in RPD, designing of connectors and pontics, etc. Given this vast amount of data being generated, dentists/prosthodontists today are faced with an overwhelming amount of information when working to rehabilitate even a single patient. AI has the potential to provide these professionals with the ability to speed up and improve their rehabilitative capabilities by helping to extract clinically relevant insights from the wealth of information available. AI technologies are making great strides in medical imaging. Studies showed that the use of AI may be able to enable earlier disease detection, while also enhancing the workflows. AI can review r vast numbers of images and then, quickly and regularly identify patterns, including variations that humans cannot. This may improve patient rehabilitative treatment outcomes. This adds to the huge potential of AI to support clinical decisions in time-critical situations or when there is a lack of expert knowledge available. AI has incredible potential for analysis and diagnosis. Most of the time and effort in a lab is spent on pre- and post-analytical processes. AI could help bring significant improvements to the workflow and operations, saving time, labor, and costs.

3.4. Error Reduction

Similarly, error reduction was enhanced with the use of AI in prosthodontics. When using traditional dental methods, there may be several issues when cementing implant prostheses, postural mistakes, directional errors, cementation faults, and occlusal or interproximal errors that might occur. With the use of AI, the possibility of errors was reduced to a minimum and the development of satisfactory functional and aesthetic prostheses was increased.

4. Discussion

In this comprehensive assessment of AI technologies utilized in prosthodontics, the predictive power and identification potential of AI for application in automated diagnostics were proven [72–74]. Due to the rapid evolution of digital technologies, in the past 2–3 years, the time frame for this investigation was set to the last four years. This systematic study was not intended to serve as a comprehensive history of AI in dentistry but rather to highlight applications for prosthetic AI. The findings demonstrated that CAD/CAM systems, implant prostheses, and studies of orofacial anatomy were just a few examples of the ways AI was used in prosthodontics. The application of AI in prosthodontics was the subject of only 24 qualifying studies. Although dynamic caries detection was the focus of artificial intelligence dental image analysis diagnosis for some time, the discipline moved to other areas of interest, demonstrating the utilization of AI technology in prosthodontics, and the large and complex field of prosthodontics in dentistry may profit from routine AI technology application [75–77].

For a prosthetic reconstruction to be effective, it must have a synoptic treatment idea, sufficient backward planning, and clean practical execution, including dental laboratory processes. Using prosthetic AI, it was possible to diagnose periodontal deficient premolars and molars with 90% and 95% accuracy, respectively [78–81]. However, because of the redundancy of the imaging characteristics and the visual field of periapical radiographs, this method cannot distinguish between early lesions or offer a conclusive diagnosis of periodontal disease.

4.1. Application of AI in Prosthodontics

Artificial intelligence (AI) uses machine-learning models to simulate human intellect and behavior. This model is based on the statistical analysis of past data and was trained using previously gathered data [82–86]. Digital data are generated at an exponential rate, which helps to train AI systems to produce more precise results. With the introduction of artificial intelligence-based technologies in prosthodontics, fundamental shifts were witnessed in their application to automatic diagnoses, predictive measurements, and classification or diagnostic tools [63]. All aspects of modern dental technology are used in prosthodontics. The digital impression with an intraoral scanner replaced the more traditional methods of impression-making. Intraoral scanners are reliable enough for everyday use, especially when only a single crown or short-span FPD is to be fabricated [64]. However, advancements in the scanning field led to its use in complete denture fabrication and maxillofacial intraoral scans.

In fixed prosthodontics, margin detection was completed using AI following an intraoral scan [65]. CAD/CAM, a common acronym for "computer-aided design/computer-aided manufacturing", is used in the creation of both permanent and removable dental prostheses [34]. An ideal crown design for a variety of circumstances may be provided by this technique using data from many actual crowns.

In recent times, many different areas of dentistry used digital tools to help patients obtain the beautiful new smiles they always wanted. These include 3D face tracking and affordable virtual 3D data hybrids such as fragmented cone beam computed tomography (CBCT), intraoral scans, and face scans. Any therapeutic action that changes a patient's smile is predicated on the virtualization of their anatomy [67]. The initial grin designs were created using simple sheet drawings made from two-dimensional printed pictures of patients [68].

Various applications of AI in prosthetic dentistry and its combination with other branches of dentistry led to a wide variety of innovative opportunities, such as the generation of occlusal morphology in crown contemplation of the opposing teeth, even in cases of wear or fracture, programmed teeth setting for dentures, or automatic framework designs for removable dental prostheses [41]. Ultimately, AI, when used as an educational tool, guides new students, graduates, and even postgraduates. AI also provides the opportunity to support less-experienced undergraduate students in their professional development [42].

4.2. Use of AI in Implant Prosthodontics

In the field of prosthodontics, both the patient and doctor need the greatest caliber of prostheses. A faultless result requires much equipment and work; however, it occasionally falls short [69]. A developing and manufacturing unit built into a computer enabled us to develop, grind, or print according to a patient's preferences while simultaneously conserving time and materials [70]. The abundance of information available on dental anatomy is frequently considered when evaluating aesthetics. This was in comparison to using a typical framework [71]. The best treatment strategy for dental implants incorporates both intraoral scanning and CBCT. AI's use of AI in implant dentistry presents the opportunity to merge the two and produce next-generation prosthetics [87]. The treatment of temporary and removable dental prostheses, the design of completing margins next to the teeth for improved longevity and alignment of the prosthesis, implant surgery, creation of maxillofacial prostheses, maintenance of ideal maxillo-mandibular relationships, and selection of teeth color for improved appearance are the primary areas of attention in prosthodontics [88]. AI has several benefits and can be applied to many treatment procedures [89]. In research by Lee J et al. [53], panoramic and periapical radiography were utilized to classify implants using convolutional neural networks (CNNs) based on AI [90,91]. Based on the results of this study, it can be said that the AI-CNN system is virtually as effective as humans in classifying implant methods [91]. Potential causes of errors include incorrect positioning, poor cementation, occlusion, and interproximal repair. An AI model was proposed by Lerner et al. [54] to lessen the likelihood of these errors. A systematic study was carried out by Takahashi et al. [92] to develop an AI framework that would categorize dental arches and utilize CNN to assist in denture manufacturing. Using computer-based autonomous learning approaches, the training dataset was categorized [93–95]. AI augmented reality reduced dental fear and improved patient satisfaction. AI will improve in organizing appointments, playing patients' favorite music and entertainment, and even in helping them to relax in their surroundings [15].

4.3. CAD/CAM and AI

Patients and clinicians anticipate high-quality prostheses in prosthodontics. Perfect output requires a lot of personnel and equipment. Computers have built-in creating and producing units that allow us to design, mill, or print custom medical prostheses for patients. The capacity of AI to evaluate and understand prostheses in the database is a significant advantage, especially when considering that new instances are added to the web on a mass scale. Dental anatomy data were used to assess aesthetics [55,82].

4.4. Maxillofacial Prostheses and AI

AI employs convolutional neural networks (CNNs) that mimic human neurons. Twelve patients with vision impairments previously tried the prosthetic eye, which was created in the United States. These AI-powered gadgets can help individuals see without surgery. AI and certain designing tools help dentists build the most beautiful prosthesis for patients, considering anthropological calculations, face dimensions, ethnicity, and patient preference. There are smart reading glasses available for the blind and visually challenged. It is an innovative voice-activated gadget that can be attached to almost any pair of glasses. It is designed primarily to help blind and visually challenged people. It can quickly read text from a book, smartphone screen, or any other surface, identify faces, work more effectively, and help its user lead an independent life [75].

The audio output can be heard using the normal microphones used in mobile phones. Skin tissue engineering is a contemporary medical practice that aims to create bioprinter biomaterial-based synthetic skin grafts. This cutting-edge approach to wound regeneration attempts to create skin replacements that work as bioactive dressings, improving the wound's functionality.

As a result, the primary functions of tissue-engineered skin grafts are to give oxygen, prevent the wound from becoming dehydrated, promote healing, and guard against infections. Artificial skin grafts can serve as temporary wound coverings or as long-term skin replacements. Artificial olfaction played a crucial role in mimicking the human olfactory structure for around 40 years; artificial olfactory systems have captivated scientists. According to research, four chemical sensors with overlapping selection patterns may differ between various odors. The categorization, identification, and recognition of scents are all based on the signal combination pattern that emerges in the ensemble of each receptor. As an example of an artificial olfactory system, the electronic nose model, which mimics the human olfactory detection system utilizing a variety of electronic sensors, was developed [35].

The bionic eye was made in the United States, and a dozen people who had lost their sight tried it. AI is used in these technologies, which can help people see without having to have surgery. With this method, a smart camera on special glasses lets the user read text or recognize faces. An expert observes the data from the camera and turns it into sound. This sound is then sent to the blind person's ears through a wireless earpiece. People who have had limbs cut off may lose the ability to feel in those places. Researchers at the Federal Polytechnic School of Zurich in Zurich, Switzerland, and the California Institute of Technology in Pasadena, CA, USA, made artificial skin that changed this picture. The tissue, made of a thin, see-through layer of water and pectin, can sense changes in temperature between 5 and 50 °C [19]. Artificial olfaction is important in robotics because it mimics the human olfactory system, which can recognize different smells in a wide range of sectors, such as disease diagnosis, environmental monitoring, public safety issues, the food industry, and agricultural production [22].

4.5. Limitations of AI

The impact of artificial intelligence on society is already significant and is expected to grow as technology improves. There is always a chance that a badly spelt symbol in the algorithm may cause a major error in the operation, or that overloaded mechanisms will simply cause the system to crash. Additionally, as prosthodontics is entrusting artificial intelligence technologies with more and more crucial tasks, the results of such failures may have unanticipated and highly unfavourable effects.

There is no agreement among contemporary scholars on how to categorize the outcomes of artificial intelligence's acts and whether they produce harm. There are three key ideas: first, identifying the person who used artificial intelligence responsibly; second, identifying the programmer (software developer), if the error was caused by a mistake or failure; and third, identifying the person who owns the rights to artificial intelligence. Today, artificial intelligence technologies are still not fully understood due to the complexity of their systems, and their ability to learn by themselves and change their behaviour.

The information provided to the AI algorithms is the only way they can learn. However, if the software is given faulty or untrustworthy data, the findings may be biased. As a result, the intelligence or effectiveness of AI is only as good as the data provided.

A machine executes an algorithm, which may or may not be programmed by a person. However, flawed algorithms will provide unfair outcomes. AI is known for learning from massive datasets, finding patterns, and making data-driven judgments. Despite its speed and accuracy, the AI system cannot explain how it reached its conclusions, and there may be substantial initial investment as well as regular maintenance and repair costs associated with this new technology. Keeping up with the ever-changing demands of prosthodontics requires that AI software receives regular updates.

4.6. Ethical Considerations with AI

AI development should ensure that these programs do not damage people while retaining the morality of computers [74]. There is great potential for the use of AI in clinical practice to advance healthcare, but this potential comes with serious ethical concerns that need to be addressed. Some of today's most powerful companies in the IT sector think that AI should be used more broadly. However, there are numerous ethical and risk assessment considerations to consider before this becomes a reality. Four fundamental ethical challenges must be solved to maximize AI in healthcare: data privacy, informed permission to use data, safety and transparency, and algorithmic fairness and biases. The use of AI may render us unable to hold anyone liable for any harm caused. Machines will restrict our capacity to assign blame and take responsibility for decisions, and the threat is unclear. Healthcare AI must adapt to a constantly changing environment with frequent interruptions while upholding ethical standards to protect patients [75].

4.7. Future Scope

AI is changing everything from space research to dentistry. Biomedical diagnosis, therapy planning, patient recording, and management have several benefits. AI aids physicians and patients in every profession, and in the future, AI will produce a forecast that can be merged with human diagnosis to increase the chances of appropriate diagnostics, leading to a higher rate of correct diagnoses.

Despite the encouraging outcomes, it is still important to validate the generalizability and reliability of the provided AI models using adequate external data acquired from newly enrolled patients or gathered from other dental facilities. One of the long-term goals of

AI research in dentistry is to improve AI models to the point where they can detect early abnormalities that are undetectable to the naked eye. There will be a significant need for AI-enabled algorithms as CAD/CAM technology advances and prosthodontic implant procedures require more accuracy.

At all stages of AI development, including pre-modelling, model creation, and post-modelling, explainability can be taken into account. The majority of the research on AI explainability describes post-modelling explainability and seeks to explain an existing black-box model version [96], even though explaining decisions made by artificial intelligence systems can help provide transparency on how the model arrives at its decision. However, at the level of the dentist and specific prosthodontist, the applications of AI is concerned with prosthesis fabrication, the success of which will be tested and determined only when it is placed in the oral cavity and so, if the explainability of AI is not known to us as a prosthodontist, the ultimate effect will be negligible. Despite this, we accept that if it is known it will be beneficial to both the prosthodontist and patient, then it will be beyond the scope of the present review. It is recommended to involve this aspect of AI in future studies and reviews.

AI will improve the clinical and dental patient experience. To improve patient experience, the system will learn preferences. Improved dental patient experiences will increase appropriate oral health care, enhancing systemic health. The software will offer RPD designs for partial edentulism. Research-based, clinically proven technologies and approaches will modernize dental implant therapy. The software will assist with partial denture design in cases of incomplete edentulism. Dental implant therapy can become standardized via the use of tried-and-true technology and procedures backed by years of scientific study.

5. Conclusions

It can be concluded that the use of AI is increasing in prosthodontics and its use is enhancing the prosthodontic-driven rehabilitation of patients. AI is helpful in removable, fixed, maxillofacial, and implant prosthodontics. The functionality and acceptance of prosthodontic treatment are enhanced with the use of AI and chances of human error are reduced. It was also revealed that prosthodontic implant applications benefit the most from artificial intelligence. In addition, researchers were found to use AI to create systems for dentistry and overall health.

Author Contributions: Conceptualization A.A.F.A., A.A.A., L.I.N.A., S.C., R.S.S., S.A.B.H., M.C. and G.M.; methodology, A.A.F.A., A.A.A., L.I.N.A., S.C., R.S.S., S.A.B.H., M.C. and G.M.; software, A.A.F.A., A.A.A., L.I.N.A., S.C. and R.S.S.; validation, A.A.F.A., A.A.A., L.I.N.A., S.C. and R.S.S.; formal analysis, A.A.F.A., A.A.A., L.I.N.A., S.C. and R.S.S.; investigation A.A.F.A., A.A.A., L.I.N.A., S.C., R.S.S., S.A.B.H., M.C. and G.M.; resources, A.A.F.A., A.A.A., L.I.N.A., S.C., R.S.S., S.A.B.H., M.C. and G.M.; data curation, A.A.F.A., A.A.A., L.I.N.A., S.C., R.S.S., S.A.B.H., M.C. and G.M.; writing—original draft preparation, A.A.F.A., A.A.A., L.I.N.A., S.C. and R.S.S.; writing—review and editing, A.A.F.A., A.A.A., L.I.N.A., S.C., R.S.S., S.A.B.H., M.C. and G.M.; visualization, A.A.F.A., A.A.A., L.I.N.A., S.C. and R.S.S.; supervision, A.A.F.A., A.A.A., L.I.N.A., S.C. and R.S.S.; project administration, A.A.F.A., A.A.A., L.I.N.A., S.C., R.S.S., S.A.B.H., M.C. and G.M.; funding acquisition, A.A.F.A., A.A.A., L.I.N.A., S.C., R.S.S., S.A.B.H., M.C. and G.M. All authors have read and agreed to the published version of the manuscript.

Funding: The authors thank the Deanship of Scientific Research at King Khalid University, Abha, Saudi Arabia, for supporting the present study through (GRP/37/43).

Institutional Review Board Statement: Not applicable.

Informed Consent Statement: Not applicable.

Data Availability Statement: Data can be provided for academic purposes on request from the primary author.

Acknowledgments: We would like to express our gratitude to our Abdullah Shahrani for the continuous support for our research and also to all of our colleagues in the department who helped in finalizing this project.

Conflicts of Interest: The authors declare no conflict of interest.

References

1. Mahdi, S.S.; Battineni, G.; Khawaja, M.; Allana, R.; Siddiqui, M.K.; Agha, D. How does artificial intelligence impact digital healthcare initiatives? A review of AI applications in dental healthcare. *Int. J. Inf. Manag. Data Insights* **2023**, *3*, 100144. [CrossRef]
2. Femiano, F.; Femiano, R.; Femiano, L.; Nucci, L.; Minervini, G.; Antonelli, A.; Bennardo, F.; Barone, S.; Scotti, N.; Sorice, V.; et al. A New Combined Protocol to Treat the Dentin Hypersensitivity Associated with Non-Carious Cervical Lesions: A Randomized Controlled Trial. *Appl. Sci.* **2020**, *11*, 187. [CrossRef]
3. Evangelista, K.; de Freitas Silva, B.S.; Yamamoto-Silva, F.P.; Valladares-Neto, J.; Silva, M.A.G.; Cevidanes, L.H.S.; de Luca Canto, G.; Massignan, C. Accuracy of artificial intelligence for tooth extraction decision-making in orthodontics: A systematic review and meta-analysis. *Clin. Oral Investig.* **2022**, *26*, 6893–6905. [CrossRef] [PubMed]
4. Pareek, M.; Kaushik, B. Artificial intelligence in prosthodontics: A scoping review on current applications and future possibilities. *Int. J. Adv. Med.* **2022**, *9*, 367. [CrossRef]
5. Cicciù, M. Prosthesis: New Technological Opportunities and Innovative Biomedical Devices. *Prosthesis* **2019**, *1*, 1–2. [CrossRef]
6. Cicciù, M.; Cervino, G.; Terranova, A.; Risitano, G.; Raffaele, M.; Cucinotta, F.; Santonocito, D.; Fiorillo, L. Prosthetic and Mechanical Parameters of the Facial Bone under the Load of Different Dental Implant Shapes: A Parametric Study. *Prosthesis* **2019**, *1*, 41–53. [CrossRef]
7. Alsubaiy, E.F.; Chaturvedi, S.; Qutub, O.A.; Mously, H.A.; Zarbah, M.A.; Haralur, S.B.; Bhagat, T.V. Novel CAD-CAM zirconia coping design to enhance the aesthetics and strength for anterior PLZ crowns. *Technol. Health Care* **2021**, *29*, 1161–1171. [CrossRef]
8. Cicciù, M. Past, Present and Future of Prosthesis: 2020 Upgrade. *Prosthesis* **2021**, *3*, 9–10. [CrossRef]
9. Chaturvedi, S.; Alqahtani, N.M.; Addas, M.K.; Alfarsi, M.A. Marginal and internal fit of provisional crowns fabricated using 3D printing technology. *Technol. Health Care* **2020**, *28*, 635–642. [CrossRef]
10. Minervini, G.; Fiorillo, L.; Russo, D.; Lanza, A.; D'Amico, C.; Cervino, G.; Meto, A.; di Francesco, F. Prosthodontic Treatment in Patients with Temporomandibular Disorders and Orofacial Pain and/or Bruxism: A Review of the Literature. *Prosthesis* **2022**, *4*, 253–262. [CrossRef]
11. Abella Sans, F. Minimally Invasive Alternatives to Dental Extraction and Implant Placement. In *Minimally Invasive Approaches in Endodontic Practice*; Springer International Publishing: Cham, Switzerland, 2021; pp. 203–231.
12. Paschou, T.; Rapaccini, M.; Adrodegari, F.; Saccani, N. Digital servitization in manufacturing: A systematic literature review and research agenda. *Ind. Mark. Manag.* **2020**, *89*, 278–292. [CrossRef]
13. Naddeo, P.; Laino, L.; la Noce, M.; Piattelli, A.; de Rosa, A.; Iezzi, G.; Laino, G.; Paino, F.; Papaccio, G.; Tirino, V. Surface biocompatibility of differently textured titanium implants with mesenchymal stem cells. *Dent. Mater.* **2015**, *31*, 235–243. [CrossRef] [PubMed]
14. Shah, K.C.; Lum, M.G. Treatment planning for the single-tooth implant restoration–general considerations and the pretreatment evaluation. *J. Calif. Dent. Asso.c* **2008**, *36*, 827–834.
15. Jihed, M.; Dallel, I.; Tobji, S.; ben Amor, A. The Impact of Artificial Intelligence on Contemporary Orthodontic Treatment Planning-A Systematic Review and Meta-Analysis. *Sch. J. Dent. Sci.* **2022**, *9*, 70–87. [CrossRef]
16. Seckanovic, A.; Sehovac, M.; Spahic, L.; Ramic, I.; Mamatnazarova, N.; Pokvic, L.G.; Badnjevic, A.; Kacila, M. Review of Artificial Intelligence Application in Cardiology. In Proceedings of the 2020 9th Mediterranean Conference on Embedded Computing, MECO 2020, Budva, Montenegro, 8–11 June 2020. [CrossRef]
17. Stokes, K.; Castaldo, R.; Franzese, M.; Salvatore, M.; Fico, G.; Pokvic, L.G.; Badnjevic, A.; Pecchia, L. A machine learning model for supporting symptom-based referral and diagnosis of bronchitis and pneumonia in limited resource settings. *Biocybern. Biomed. Eng.* **2021**, *41*, 1288–1302. [CrossRef]
18. Granulo, E.; Bećar, L.; Gurbeta, L.; Badnjević, A. Telemetry system for diagnosis of asthma and chronical obstructive pulmonary disease (COPD). In *Lecture Notes of the Institute for Computer Sciences, Social-Informatics and Telecommunications Engineering, LNICST*; Springer International Publishing: Cham, Switzerland, 2016. [CrossRef]
19. Shajahan, P.A.; Raghavan, R.; Joe, N. Application of artificial intelligence in prosthodontics. *Int. J. Sci. Health Care Res.* **2021**, *6*, 57–60.
20. Saghiri, M.A.; Vakhnovetsky, J.; Nadershahi, N. Scoping review of artificial intelligence and immersive digital tools in dental education. *J. Dent. Educ.* **2022**, *86*, 736–750. [CrossRef] [PubMed]
21. Cicciù, M.; Fiorillo, L.; D'Amico, C.; Gambino, D.; Amantia, E.M.; Laino, L.; Crimi, S.; Campagna, P.; Bianchi, A.; Herford, A.S.; et al. 3D Digital Impression Systems Compared with Traditional Techniques in Dentistry: A Recent Data Systematic Review. *Materials* **2020**, *13*, 1982. [CrossRef]
22. Agrawal, P.; Nikhade, P. Artificial Intelligence in Dentistry: Past, Present, and Future. *Cureus* **2022**, *14*, e27405. [CrossRef]
23. Ametrano, G.; D'Antò, V.; di Caprio, M.P.; Simeone, M.; Rengo, S.; Spagnuolo, G. Effects of sodium hypochlorite and ethylenediaminetetraacetic acid on rotary nickel-titanium instruments evaluated using atomic force microscopy. *Int. Endod. J.* **2011**, *44*, 203–209. [CrossRef]

24. Krifka, S.; Petzel, C.; Bolay, C.; Hiller, K.-A.; Spagnuolo, G.; Schmalz, G.; Schweikl, H. Activation of stress-regulated transcription factors by triethylene glycol dimethacrylate monomer. *Biomaterials* **2011**, *32*, 1787–1795. [CrossRef] [PubMed]
25. Chakraborty, T.; Jamal, R.F.; Battineni, G.; Teja, K.V.; Marto, C.M.; Spagnuolo, G. A review of prolonged post-COVID-19 symptoms and their implications on dental management. *Int. J. Environ. Res. Public Health* **2021**, *18*, 5131. [CrossRef] [PubMed]
26. Rengo, C.; Goracci, C.; Ametrano, G.; Chieffi, N.; Spagnuolo, G.; Rengo, S.; Ferrari, M. Marginal leakage of class v composite restorations assessed using microcomputed tomography and scanning electron microscope. *Oper. Dent.* **2015**, *40*, 440–448. [CrossRef] [PubMed]
27. Spagnuolo, G.; Sorrentino, R. The Role of Digital Devices in Dentistry: Clinical Trends and Scientific Evidences. *J. Clin. Med.* **2020**, *9*, 1692. [CrossRef] [PubMed]
28. Kochhar, A.S.; Sidhu, M.S.; Prabhakar, M.; Bhasin, R.; Kochhar, G.K.; Dadlani, H.; Spagnuolo, G.; Mehta, V.V. Intra- and Interobserver Reliability of Bone Volume Estimation Using OsiriX Software in Patients with Cleft Lip and Palate Using Cone Beam Computed Tomography. *Dent. J.* **2021**, *9*, 14. [CrossRef]
29. Abdinian, M.; Moshkforoush, S.; Hemati, H.; Soltani, P.; Moshkforoushan, M.; Spagnuolo, G. Comparison of Cone Beam Computed Tomography and Digital Radiography in Detecting Separated Endodontic Files and Strip Perforation. *Appl. Sci.* **2020**, *10*, 8726. [CrossRef]
30. Spagnuolo, G. Cone-Beam Computed Tomography and the Related Scientific Evidence. *Appl. Sci.* **2022**, *12*, 7140. [CrossRef]
31. Rengo, C.; Spagnuolo, G.; Ametrano, G.; Juloski, J.; Rengo, S.; Ferrari, M. Micro-computerized tomographic analysis of premolars restored with oval and circular posts. *Clin. Oral Investig.* **2014**, *18*, 571–578. [CrossRef]
32. Bernauer, S.A.; Zitzmann, N.U.; Joda, T. The Use and Performance of Artificial Intelligence in Prosthodontics: A Systematic Review. *Sensors* **2021**, *21*, 6628. [CrossRef]
33. Thurzo, A.; Urbanová, W.; Novák, B.; Czako, L.; Siebert, T.; Stano, P.; Mareková, S.; Fountoulaki, G.; Kosnáčová, H.; Varga, I. Where Is the Artificial Intelligence Applied in Dentistry? Systematic Review and Literature Analysis. *Healthcare* **2022**, *10*, 1269. [CrossRef]
34. Moher, D.; Liberati, A.; Tetzlaff, J.; Altman, D.G.; Group, P. Preferred reporting items for systematic reviews and meta-analyses: The PRISMA statement. *PLoS Med.* **2009**, *6*, e1000097. [CrossRef] [PubMed]
35. Grischke, J.; Johannsmeier, L.; Eich, L.; Griga, L.; Haddadin, S. Dentronics: Towards robotics and artificial intelligence in dentistry. *Dent. Mater.* **2020**, *36*, 765–778. [CrossRef] [PubMed]
36. Lahoud, P.; EzEldeen, M.; Beznik, T.; Willems, H.; Leite, A.; Van Gerven, A.; Jacobs, R. Artificial Intelligence for fast and accurate 3-dimensional tooth segmentation on cone-beam computed tomography. *J. Endod.* **2021**, *47*, 827–835. [CrossRef] [PubMed]
37. Farook, T.H.; Jamayet, N.B.; Abdullah, J.Y.; Alam, M.K. Machine learning and intelligent diagnostics in dental and Orofacial Pain Management: A systematic review. *Pain Res. Manag.* **2021**, *2021*, 6659133. [CrossRef] [PubMed]
38. Talpur, S.; Azim, F.; Rashid, M.; Syed, S.A.; Talpur, B.A.; Khan, S.J. Use of different machine learning algorithms for diagnosis of dental caries. *J. Healthc. Eng.* **2022**, *2022*, 5032435. [CrossRef]
39. Limones, A.; Celemín-Viñuela, A.; Romeo-Rubio, M.; Castillo-Oyagüe, R.; Gómez-Polo, M.; Martínez Vázquez de Parga, J.A. Outcome measurements and quality of randomized controlled clinical trials of tooth-supported fixed dental prostheses: A systematic review and qualitative analysis. *J. Prosthet. Dent.* **2022**, *S0022-3913*, 00282-7. [CrossRef]
40. van Riet, T.C.T.; Chin Jen Sem, K.T.H.; Ho, J.-P.T.F.; Spijker, R.; Kober, J.; de Lange, J. Robot Technology in Dentistry: Part One of a Systematic Review: Literature Characteristics. *Dent. Mater.* **2021**, *37*, 1217–1226. [CrossRef]
41. Delgado, A.H.S.; Sauro, S.; Lima, A.F.; Loguercio, A.D.; Della Bona, A.; Mazzoni, A.; Collares, F.M.; Staxrud, F.; Ferracane, J.; Tsoi, J.; et al. Risk of bias tool and guideline to support reporting of pre-clinical Dental Materials Research and assessment of Systematic Reviews. *J. Dent.* **2022**, *127*, 104350. [CrossRef]
42. Kirubarajan, A.; Young, D.; Khan, S.; Crasto, N.; Sobel, M.; Sussman, D. Artificial Intelligence and Surgical Education: A systematic scoping review of interventions. *J. Surg. Educ.* **2022**, *79*, 500–515. [CrossRef]
43. Afrashtehfar, K.I.; Alnakeb, N.A.; Assery, M.K. Accuracy of intraoral scanners versus traditional impressions: A Rapid Umbrella Review. *J. Evid. Based Dent. Pract.* **2022**, *22*, 101719. [CrossRef]
44. Albanchez-González, M.I.; Brinkmann, J.C.-B.; Peláez-Rico, J.; López-Suárez, C.; Rodríguez-Alonso, V.; Suárez-García, M.J. Accuracy of digital dental implant impression using intraoral scanners compared with conventional impression techniques: A systematic review of in vitro studies. *Int. J. Environ. Res. Public Health* **2022**, *19*, 2026. [CrossRef]
45. Mai, H.Y.; Mai, H.N.; Kim, H.J.; Lee, J.; Lee, D.H. Accuracy of removable partial denture metal frameworks fabricated by computer-aided design/computer-aided manufacturing method: A systematic review and meta-analysis. *J. Evid. Based Dent. Pract.* **2022**, *22*, 101681. [CrossRef]
46. Touati, R.; Sailer, I.; Marchand, L.; Ducret, M.; Strasding, M. Communication tools and patient satisfaction: A scoping review. *J. Esthet. Restor. Dent.* **2021**, *34*, 104–116. [CrossRef] [PubMed]
47. Adel, S.; Zaher, A.; El Harouni, N.; Venugopal, A.; Premjani, P.; Vaid, N. Robotic applications in orthodontics: Changes in Contemporary Clinical Care. *BioMed Res. Int.* **2021**, *2021*, 9954615. [CrossRef]
48. Tian, Y.; Chen, C.X.; Xu, X.; Wang, J.; Hou, X.; Li, K.; Lu, X.; Shi, H.Y.; Lee, E.-S.; Jiang, H.B. Review of 3D printing in Dentistry: Technologies, affecting factors, and applications. *Scanning* **2021**, *2021*, 9950131. [CrossRef]
49. Nguyen, T.M.; Tonmukayakul, U.; Le, L.K.-D.; Calache, H.; Mihalopoulos, C. Economic evaluation of preventive interventions for dental caries and periodontitis: A systematic review. *Appl. Health Econ. Health Policy* **2022**, *21*, 53–70. [CrossRef]

50. Costa, R.T.F.; de Oliveira Limirio, J.P.J.; do Egito Vasconcelos, B.C.; Pellizzer, E.P.; de Moraes, S.L.D. Rehabilitation with dental prostheses and its influence on brain activity: A systematic review. *J. Prosthet. Dent.* **2022**, *S0022-3913*, 00090-7. [CrossRef] [PubMed]
51. Sridharan, K.; Alsobaiei, M.; Sivaramakrishnan, G.; AlSulaiti, F. Detection of caries under fixed prosthodontic restorations using cone-beam CT: A meta-analysis. *Int. J. Prosthodont. Restor. Dent.* **2020**, *10*, 170–175. [CrossRef]
52. Abad-Coronel, C.; Valdiviezo, O.P.; Naranjo, O.B. Intraoral scanning devices are used in fixed prosthodontics. *Acta Sci. Dent. Sci.* **2019**, *7*, 44–51.
53. Lee, J.H.; Jeong, S.N. Efficacy of deep convolutional neural network algorithm for the identification and classification of dental implant systems, using panoramic and periapical radiographs: A pilot study. *Medicine* **2020**, *99*, e20787. [CrossRef] [PubMed]
54. Lerner, H.; Mouhyi, J.; Admakin, O.; Mangano, F. Artificial intelligence in fixed implant prosthodontics: A retrospective study of 106 implant-supported monolithic zirconia crowns inserted in the posterior jaws of 90 patients. *BMC Oral Health* **2020**, *20*, 80. [CrossRef]
55. Yamaguchi, S.; Lee, C.; Karaer, O.; Ban, S.; Mine, A.; Imazato, S. Predicting the Debonding of CAD/CAM Composite Resin Crowns with AI. *J. Dent. Res.* **2019**, *98*, 1234–1238. [CrossRef]
56. Lee, J.H.; Kim, D.H.; Jeong, S.N.; Choi, S.H. Detection and diagnosis of dental caries using a deep learning-based convolutional neural network algorithm. *J. Dent.* **2018**, *77*, 106–111. [CrossRef]
57. Lee, J.H.; Kim, D.H.; Jeong, S.N.; Choi, S.H. Diagnosis and prediction of periodontally compromised teeth using a deep learning-based convolutional neural network algorithm. *J. Periodontal Implant Sci.* **2018**, *48*, 114–123. [CrossRef] [PubMed]
58. Raith, S.; Vogel, E.P.; Anees, N.; Keul, C.; Güth, J.F.; Edelhoff, D.; Fischer, H. Artificial Neural Networks as a powerful numerical tool to classify specific features of a tooth based on 3D scan data. *Comput. Biol. Med.* **2017**, *80*, 65–76. [CrossRef] [PubMed]
59. Wei, J.; Peng, M.; Li, Q.; Wang, Y. Evaluation of a Novel Computer Color Matching System Based on the Improved Back-Propagation Neural Network Model. *J. Prosthodont.* **2018**, *27*, 775–783. [CrossRef]
60. Ahmed, N.; Abbasi, M.S.; Zuberi, F.; Qamar, W.; Halim, M.S.B.; Maqsood, A.; Alam, M.K. Artificial Intelligence Techniques: Analysis, Application, and Outcome in Dentistry—A Systematic Review. *Biomed. Res. Int.* **2021**, *2021*, 9751564. [CrossRef] [PubMed]
61. Khanagar, S.B.; Al-ehaideb, A.; Maganur, P.C.; Vishwanathaiah, S.; Patil, S.; Baeshen, H.A.; Sarode, S.C.; Bhandi, S. Developments, application, and performance of artificial intelligence in dentistry–A systematic review. *J. Dent. Sci.* **2021**, *16*, 508–522. [CrossRef] [PubMed]
62. Fontenele, R.C.; Gerhardt, M.d.N.; Pinto, J.C.; van Gerven, A.; Willems, H.; Jacobs, R.; Freitas, D.Q. Influence of dental fillings and tooth type on the performance of a novel artificial intelligence-driven tool for automatic tooth segmentation on CBCT images—A validation study. *J. Dent.* **2022**, *119*, 104069. [CrossRef] [PubMed]
63. Abouzeid, H.L.; Chaturvedi, S.; Abdelaziz, K.M.; Alzahrani, F.A.; AlQarni, A.A.S.; Alqahtani, N.M. Role of Robotics and Artificial Intelligence in Oral Health and Preventive Dentistry-Knowledge, Perception and Attitude of Dentists. *Oral Health Prev. Dent.* **2021**, *19*, 353–363.
64. Amornvit, P.; Rokaya, D.; Sanohkan, S. Comparison of Accuracy of Current Ten Intraoral Scanners. *Biomed Res. Int.* **2021**, *2021*, 2673040. [CrossRef]
65. Cabanes-Gumbau, G.; Palma, J.C.; Kois, J.C.; Revilla-León, M. Transferring the tooth preparation finish line on intraoral digital scans to dental software programs: A dental technique. *J. Prosthet. Dent.* **2022**, *S0022-3913*, 00582-5. [CrossRef]
66. van der Meer, W.J.; Andriessen, F.S.; Wismeijer, D.; Ren, Y. Application of intra-oral dental scanners in the digital workflow of implantology. *PLoS ONE* **2012**, *7*, e43312. [CrossRef]
67. Rekow, E.D. Digital dentistry: The new state of the art—Is it disruptive or destructive? *Dent. Mater.* **2020**, *36*, 9–24. [CrossRef] [PubMed]
68. Jreige, C.S.; Kimura, R.N.; Segundo, Â.R.T.C.; Coachman, C.; Sesma, N. Esthetic treatment planning with digital animation of the smile dynamics: A technique to create a 4-dimensional virtual patient. *J. Prosthet. Dent.* **2022**, *128*, 130–138. [CrossRef]
69. Lahoud, P.; Jacobs, R.; Boisse, P.; EzEldeen, M.; Ducret, M.; Richert, R. Precision medicine using patient-specific modelling: State of the art and perspectives in dental practice. *Clin. Oral Investig.* **2022**, *26*, 5117–5128. [CrossRef] [PubMed]
70. Shen, K.; Huang, C.; Lin, Y.; Du, J.; Chen, F.; Kabasawa, Y.; Chen, C.; Huang, H. Effects of artificial intelligence-assisted dental monitoring intervention in patients with periodontitis: A randomized controlled trial. *J. Clin. Periodontol.* **2022**, *49*, 988–998. [CrossRef]
71. Javaid, M.; Haleem, A.; Pratap Singh, R.; Suman, R. Pedagogy and innovative care tenets in COVID-19 pandemic: An enhancive way through Dentistry 4.0. *Sens. Int.* **2021**, *2*, 100118. [CrossRef] [PubMed]
72. Hegde, S.; Ajila, V.; Zhu, W.; Zeng, C. Artificial intelligence in early diagnosis and prevention of oral cancer. *Asia-Pac. J. Oncol. Nurs.* **2022**, *9*, 100133. [CrossRef] [PubMed]
73. Thomsen, K.; Iversen, L.; Titlestad, T.L.; Winther, O. Systematic review of machine learning for diagnosis and prognosis in dermatology. *J. Dermatol. Treat.* **2020**, *31*, 496–510. [CrossRef] [PubMed]
74. Monill-González, A.; Rovira-Calatayud, L.; d'Oliveira, N.G.; Ustrell-Torrent, J.M. Artificial intelligence in orthodontics: Where are we now? A scoping review. *Orthod. Craniofac. Res.* **2021**, *24*, 6–15. [CrossRef]
75. Mureșanu, S.; Almășan, O.; Hedeșiu, M.; Dioșan, L.; Dinu, C.; Jacobs, R. Artificial intelligence models for clinical usage in dentistry with a focus on dentomaxillofacial CBCT: A systematic review. *Oral Radiol.* **2023**, *39*, 18–40. [CrossRef] [PubMed]

76. Crompton, H.; Jones, M.V.; Burke, D. Affordances and challenges of artificial intelligence in K-12 education: A systematic review. *J. Res. Technol. Educ.* **2022**, 1–21. [CrossRef]
77. Mohammad-Rahimi, H.; Nadimi, M.; Rohban, M.H.; Shamsoddin, E.; Lee, V.Y.; Motamedian, S.R. Machine learning and orthodontics, current trends and the future opportunities: A scoping review. *Am. J. Orthod. Dentofac. Orthop.* **2021**, *160*, 170–192.e4. [CrossRef]
78. Khanagar, S.B.; Vishwanathaiah, S.; Naik, S.; Al-Kheraif, A.A.; Devang Divakar, D.; Sarode, S.C.; Bhandi, S.; Patil, S. Application and performance of artificial intelligence technology in forensic odontology-A systematic review. *Leg. Med.* **2021**, *48*, 101826. [CrossRef]
79. Khanagar, S.B.; Al-Ehaideb, A.; Vishwanathaiah, S.; Maganur, P.C.; Patil, S.; Naik, S.; Baeshen, H.A.; Sarode, S.S. Scope and performance of artificial intelligence technology in orthodontic diagnosis, treatment planning, and clinical decision-making—A systematic review. *J. Dent. Sci.* **2021**, *16*, 482–492. [CrossRef] [PubMed]
80. Revilla-León, M.; Gómez-Polo, M.; Barmak, A.B.; Inam, W.; Kan, J.Y.K.; Kois, J.C.; Akal, O. Artificial intelligence models for diagnosing gingivitis and periodontal disease: A systematic review. *J. Prosthet. Dent.* **2022**, *S0022-3913*, 00075-0. [CrossRef]
81. Besse, P.; del Barrio, E.; Gordaliza, P.; Loubes, J.-M.; Risser, L. A Survey of Bias in Machine Learning Through the Prism of Statistical Parity. *Am. Stat.* **2022**, *76*, 188–198. [CrossRef]
82. Rajput, G.; Ahmed, S.; Chaturvedi, S.; Addas, M.K.; Bhagat, T.V.; Gurumurthy, V.; Alqahtani, S.M.; Alobaid, M.A.; Alsubaiy, E.F.; Gupta, K.; et al. Comparison of Microleakage in Nanocomposite and Amalgam as a Crown Foundation Material Luted with Different Luting Cements under CAD-CAM Milled Metal Crowns: An In Vitro Microscopic Study. *Polymers* **2022**, *14*, 2609.
83. Nair, U.P.; Shivamurthy, R.; Nagate, R.R.; Chaturvedi, S.; Al-Qahtani, S.M.; al Magbol, M.; Gokhale, S.T.; Tikare, S.; Chaturvedi, M. Effect of Injectable Platelet-Rich Fibrin with a Nano-Hydroxyapatite Bone Graft on the Treatment of a Grade II Furcation Defect. *Bioengineering* **2022**, *9*, 602. [CrossRef] [PubMed]
84. Mishra, S.; Chaturvedi, S.; Ali, M.; Pandey, K.K.; Alqahtani, N.M.; Alfarsi, M.A.; Addas, M.K.; Vaddamanu, S.K.; Ahmari, N.M.; Alqahtani, S.M.; et al. Dimensional Stability of Light-Activated Urethane Dimethacrylate Denture Base Resins. *Polymers* **2023**, *15*, 744. [CrossRef]
85. Rai, J.J.; Chaturvedi, S.; Gokhale, S.T.; Nagate, R.R.; Al-Qahtani, S.M.; Magbol, A.M.; Bavabeedu, S.S.; Elagib, M.F.A.; Venkataram, V.; Chaturvedi, M. Effectiveness of a Single Chair Side Application of NovaMin®[Calcium Sodium Phosphosilicate] in the Treatment of Dentine Hypersensitivity following Ultrasonic Scaling—A Randomized Controlled Trial. *Materials* **2023**, *16*, 1329. [CrossRef] [PubMed]
86. Mittal, P.; Gokhale, S.T.; Manjunath, S.; Al-Qahtani, S.M.; Magbol, A.M.; Nagate, R.R.; Tikare, S.; Chaturvedi, S.; Agarwal, A.; Venkataram, V. Comparative Evaluation of Locally Administered 2% Gel Fabricated from Lemongrass Polymer and 10% Doxycycline Hyclate Gel as an Adjunct to Scaling and Root Planing in the Treatment of Chronic Periodontitis—A Randomized Controlled Trial. *Polymers* **2022**, *14*, 2766. [CrossRef]
87. Chen, Y.-W.; Stanley, K.; Att, W. Artificial intelligence in dentistry: Current applications and future perspectives. *Quintessence Int.* **2020**, *51*, 248–257. [PubMed]
88. Bindushree, V.; Sameen, R.J.; Vasudevan, V.; Shrihari, T.G.; Devaraju, D.; Mathew, N.S. Artificial Intelligence: In Modern Dentistry. *J. Dent. Res. Rev.* **2020**, *7*, 27.
89. Revilla-León, M.; Gómez-Polo, M.; Vyas, S.; Barmak, A.B.; Gallucci, G.O.; Att, W.; Özcan, M.; Krishnamurthy, V.R. Artificial intelligence models for tooth-supported fixed and removable prosthodontics: A systematic review. *J. Prosthet. Dent.* **2021**, *129*, 276–292. [CrossRef] [PubMed]
90. Schwendicke, F.; Singh, T.; Lee, J.-H.; Gaudin, R.; Chaurasia, A.; Wiegand, T.; Uribe, S.; Krois, J. Artificial intelligence in dental research: Checklist for authors, reviewers, readers. *J. Dent.* **2021**, *107*, 103610. [CrossRef] [PubMed]
91. Jose, A.A.; Sawhney, H.; Jose, C.M.; Center, G.D. Artificial intelligence: The immeasurable limits in pediatric dentistry. *Int. J. Early Child.* **2020**, *16*, 300–309. [CrossRef]
92. Takahashi, T.; Nozaki, K.; Gonda, T.; Ikebe, K. A system for designing removable partial dentures using artificial intelligence. Part 1. Classification of partially edentulous arches using a convolutional neural network. *J. Prosthodont. Res.* **2021**, *65*, 115–118. [CrossRef] [PubMed]
93. Revilla-León, M.; Gómez-Polo, M.; Vyas, S.; Barmak, B.A.; Galluci, G.O.; Att, W.; Krishnamurthy, V.R. Artificial intelligence applications in implant dentistry: A systematic review. *J. Prosthet. Dent.* **2021**, *129*, 293–300. [CrossRef] [PubMed]
94. Asiri, A.F.; Altuwalah, A.S. The role of neural artificial intelligence for diagnosis and treatment planning in endodontics: A qualitative review. *Saudi Dent. J.* **2022**, *34*, 270–281. [CrossRef] [PubMed]
95. Mohaideen, K.; Negi, A.; Verma, D.K.; Kumar, N.; Sennimalai, K.; Negi, A. Applications of artificial intelligence and machine learning in orthognathic surgery: A scoping review. *J. Stomatol. Oral Maxillofac. Surg.* **2022**, *123*, e962–e972. [CrossRef] [PubMed]
96. Holzinger, A.; Langs, G.; Denk, H.; Zatloukal, K.; Müller, H. Causability and explainability of artificial intelligence in medicine. *Wiley Interdiscip. Rev. Data Min. Knowl. Discov.* **2019**, *9*, e1312. [CrossRef] [PubMed]

applied sciences

MDPI

Systematic Review

Acceleration Techniques for Teeth Movements in Extractive Orthodontic Therapy

Gianna Dipalma [1,†], Assunta Patano [1,†], Irene Ferrara [1,†], Fabio Viapiano [1], Anna Netti [1], Sabino Ceci [1], Daniela Azzollini [1], Anna Maria Ciocia [1], Giuseppina Malcangi [1], Alessio Danilo Inchingolo [1], Francesco Inchingolo [1,*], Andrea Palermo [2] and Angelo Michele Inchingolo [1]

1 Department of Interdisciplinary Medicine, School of Medicine, University of Bari "Aldo Moro", 70124 Bari, Italy; giannadipalma@tiscali.it (G.D.); assuntapatano@gmail.com (A.P.); ire.ferra3@gmail.com (I.F.); viapianofabio96@gmail.com (F.V.); annanetti@inwind.it (A.N.); s.ceci@studenti.uniba.it (S.C.); daniela.azzollini93@gmail.com (D.A.); anna.ciocia1@gmail.com (A.M.C.); giuseppinamalcangi@libero.it (G.M.); ad.inchingolo@libero.it (A.D.I.); angeloinchingolo@gmail.com (A.M.I.)

2 Implant Dentistry College of Medicine and Dentistry, Birmingham B4 6BN, UK; andrea.palermo2004@libero.it

* Correspondence: francesco.inchingolo@uniba.it; Tel.: +39-331-211-1104

† These authors contributed equally to this work as first authors.

Abstract: For a world that is constantly trying to speed up every procedure while obtaining the maximum result, traditional orthodontics have the biological limitation of using light and constant forces that allow tooth movement in a time frame that is only sometimes short. The treatment time could be lengthened if surgical procedures are programmed in the plan. Methods to accelerate tooth movement and reduce the duration of treatment while minimising complications are investigated and reported in the dental literature (e.g., low-level laser therapy, corticotomy, and micro-osteoperforations). This systematic review aims to analyse and summarise the strategies for quickening orthodontic movement during extraction orthodontic treatment, including any potential drawbacks or adverse consequences. The review will evaluate each approach's effectiveness, safety, and evidence quality, compare their benefits and disadvantages, and analyse the implications for clinical practice and future research. Pubmed, Science Direct, Scopus, and Web of Science were searched using the keywords "acceleration" AND "dental movement" AND "orthodontic" between 1 April 2003 and 1 April 2023. After carefully scanning the study findings, forty-four publications were chosen for the systematic review. Most therapies discussed and provided in the literature seem promising and successful in enhancing orthodontic treatments. The success of operations like corticotomies, piezo-incisions, micro-osteoperforations, osteogenic distraction, low-level laser therapy, the administration of pharmacological treatments, and infiltrations with PRF and PRP were statistically significant and appear to be promising and effective in optimising orthodontic treatments. These strategies expedite treatment and enhance the patient experience, potentially broadening orthodontic appeal and minimising issues like cavities and enamel demineralisation. Further studies, with larger samples and standardised treatment protocols, are needed to investigate the efficacy of these tooth movement acceleration modalities.

Keywords: orthodontic dental movement; acceleration tooth movement; corticotomies; low-level laser therapy; osteogenic distraction; micro-osteoperforation; piezoincision; canine retraction; extractive treatment

Citation: Dipalma, G.; Patano, A.; Ferrara, I.; Viapiano, F.; Netti, A.; Ceci, S.; Azzollini, D.; Ciocia, A.M.; Malcangi, G.; Inchingolo, A.D.; et al. Acceleration Techniques for Teeth Movements in Extractive Orthodontic Therapy. *Appl. Sci.* **2023**, *13*, 9159. https://doi.org/10.3390/app13179759

Academic Editors: Giuseppe Minervini and Rocco Franco

Received: 13 July 2023
Revised: 24 August 2023
Accepted: 26 August 2023
Published: 29 August 2023

1. Introduction

Orthodontics is a fundamental speciality of dentistry that deals with diagnosing, preventing, and treating dental and facial abnormalities. Orthodontic treatment aims to improve patients' aesthetics, function, and oral health by aligning teeth and correcting skeletal discrepancies. However, traditional orthodontic treatments can take considerable time, often between 1 and 3 years, depending on the case's complexity. This long duration of therapy may discourage some patients from seeking orthodontic intervention and expose patients to possible complications, such as enamel demineralisation, caries, and gingival recession [1–3].

In recent years, there has been a growing interest in researching methods to accelerate orthodontic tooth movement (OTM) and reduce the duration of treatment. Several approaches have been proposed and studied, including corticotomies, micro-osteoperforations (MOPs), vibration therapy (VT), low-level pulsed light therapy (LLLT), platelet-rich plasma (PRP) and platelet-rich fibrinogen (PRF), drug therapy, and dentoalveolar distraction (DAD). Despite the growing body of literature, there is still no clear consensus on which methods are most effective, safe, and feasible in the clinical setting [4–7].

This systematic review aims to examine and synthesise the evidence regarding different methods of accelerating OTM in extractive orthodontic treatment. A fixed orthodontic procedure lasts two to three years, and canine retraction is a crucial and time-consuming step in fixed orthodontic treatment for individuals who have had their premolars extracted. The canine retraction process uses traditional techniques at an average rate of 0.5 to 1 mm monthly. As a result, canine retraction alone requires 5 to 9 months and raises the risk of caries, external root resorption, and decreased patient participation. Therefore, making an effort to accelerate OTM and reduce the duration of treatment can be very helpful for improving future orthodontic treatment [8,9].

This review will evaluate the efficacy, safety, and quality of the evidence for each approach, compare the advantages and disadvantages of the various methods, and discuss the implications for clinical practice and future research. Through this analysis, we aim to provide a solid foundation for orthodontists and researchers to improve orthodontic treatments further and increase patient satisfaction.

It is important to note that not all methods are suitable for all patients and should be carefully evaluated and discussed with the orthodontist before proceeding. Some of the most used methods of accelerating OTM include:

- Corticotomy is the execution of small incisions in the alveolar bone surrounding the teeth to facilitate their movement (Figure 1) [1], using several different techniques (chisel and hammer, piezosurgery, etc.) [4]. The goal is to stimulate local biological response and bone remodelling without damaging the surrounding tissues.
- MOPs are small perforations in the alveolar bone around the teeth, obtained using miniscrews or fine needles [8]. The drilling process stimulates the local inflammatory response and accelerates bone remodelling, allowing for faster and more efficient OTM (Figure 1B) [5]. MOPs can be performed safely and with minor patient morbidity, reducing orthodontic TT [8,10].

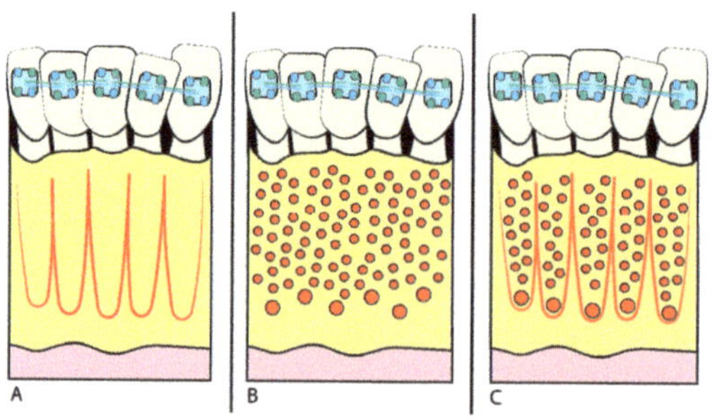

Figure 1. (**A**) Corticotomies; (**B**) MOPs; (**C**) combination of both techniques.

- VT devices use low-frequency vibrations to stimulate bone remodelling, reduce TT [11], reduce pain and discomfort, and improve the overall patient experience [6]. Application time is around 20 min per day [12].
- LLLT uses low-level pulsed light to stimulate tissue healing and bone remodelling by penetrating soft tissue without causing thermal damage [13]. This can help reduce

TT and improve patient comfort, promoting faster recovery and reducing the risk of complications [14].
- PRP and PRF are autologous platelet concentrates derived from the patient's blood [15], applied locally during orthodontic procedures to accelerate healing and bone remodelling [16].
- In some cases, nonsteroidal anti-inflammatory drugs (NSAIDs) or drugs-modulating calcium and phosphorus metabolism can also facilitate OTM and reduce pain [17].
- DAD involves a device constantly forcing the teeth to stimulate bone remodelling [18]. Osteogenic distraction is often used to treat severe skeletal discrepancies and requires close collaboration between the orthodontist and the oral-maxillofacial surgeon [19,20]. The process creates a controlled fracture in the bone, followed by applying a distraction device to lengthen the bone over time, forming new bone. This allows significant corrections of skeletal deformities and functional and aesthetic improvements [18]. Although this technique can be highly effective, it is associated with an increased risk of complications and requires careful patient management throughout the treatment process [21].

2. Materials and Methods

2.1. Protocol

The Preferred Reporting Items for Systematic Reviews and Meta-Analyses (PRISMA) guidelines were used in this systematic review (PROSPERO registration code ID 446071) [22].

2.2. Information Sources and Search Strategy

The qualifying criteria were developed using the PICOS (population, intervention, comparison, outcomes, and study design) framework. Pubmed, Science Direct, Scopus, and Web of Science databases were searched from 1 April 2003 up to 1 April 2023, using the keywords "acceleration" AND "dental movement" AND "orthodontic"(Table 1).

Table 1. Database search indicator.

Articles screening strategy	(Keywords: acceleration) AND (dental movement) AND (orthodontic) Boolean Indicators: ("A" AND "B") Timespan: from 1 April 2003 to 1 April 2023 Electronic Database: Pubmed, Science Direct, Scopus, and Web of Science

2.3. Eligibility Criteria

This research studies the dental movement acceleration strategies in extractive orthodontic treatment. Articles that met several criteria were included: (1) the study design selected was Randomised Clinical Trials (RCT), a case series with more than 5 case reports, clinical trials (CT), retrospective studies (R), and prospective studies (P); (2) participants were young adult and adult patients with permanent dentition; (3) patients were treated via extraction of the 1st or 2nd premolars in the upper or lower jaw; (4) extractive orthodontic treatment used included one strategy of dental movement acceleration (corticotomies, MOPs, VT, LLLT, drug therapy, and DAD); (5) the language selected was English; (6) only full-text was available.

Studies characterised by one of the following exclusion criteria were excluded: (1) the study design excluded was reviews, letters, or comments; case series with less than five case reports; case reports; in vivo and in vitro studies; (2) participants were animal models or dry skulls studies; (3) no-extractive orthodontic treatment; (4) the acceleration of the dental movement was facilitated using orthodontic miniscrews.

2.4. Synthesis Methods

The study data was selected by analysing the study design, number of patients, average age, dental acceleration technique employed, type of orthodontic treatment, and outcomes (Table 2).

Table 2. Descriptive summary of item selection.

Author (Year)	Study Design	Number of Patients	Average Age (Years)	Dental Movement Acceleration Techniques	Orthodontic Treatment	Outcomes
Abbas (2016) [23]	RCT	20	15–25	Piezocision/control group (CGr); corticotomy/CGr	Roth prescription brackets; closed coils NiTi springs 150 g force	Orthodontics supported via corticotomies and piezocision is 1.5–2 times faster than traditional orthodontics.
Addanki (2017) [24]	RCT	16	20–40	Buccal and palatal bur corticotomy/buccal bur corticotomy (control)	SW brackets	No difference between the two groups.
Aksakalli (2015) [7]	RCT	10	16.3 ± 2.4	Piezocision/corticocision (blade 15)	Roth prescription brackets; elastomeric chain 150 g force and medium anchorage (transpalatal arch)	Movement in the side undergoing piezocision is twice as fast as in the CGr.
Al Imam (2019) [25]	RCT	42	19.15	Piezocision	MBT prescription brackets; NiTi coil springs 150 g; medium anchorage (transpalatal arch)	The incisor retraction time in the experimental group has decreased by 27%.
Alfawal (2018) [26]	RCT	36	15–27	Piezocision/CGr; laser-assisted flapless corticotomy (LAFC) ER: YAG laser	MBT prescription brackets; NiTi closed coil spring 150 g force	The experimental side had a higher rate of OTM in the first and second months and a 25% reduction in overall canine retraction duration.
Angel et al. (2022) [27]	RCT	10	16–24	Injection of PRP	Roth prescription brackets; medium anchorage (Nance palatal button)	Movement occurred 35% more increased on the i-PRP side than on the CGr.
Arumughan et Al. (2018) [28]	RCT	12		LLLT: 810 nm wavelength laser (100 mW power, continuous wave).	MBT prescription brackets; NiTi closed-coil spring 150 g force	LLLT speeds OTM by 12.555% compared to the conventional retraction approach.
Attri et al. (2018) [29]	2-arm parallel RCT	60	13–20	MOPs	MBT prescription brackets	Increased OTM with MOPs.
Baeshen (2020) [30]	CT with the split-mouth	20	16 ± 2.8	Partial corticotomy	SW brackets; elastomeric chain 150 g force; medium anchorage (transpalatal arch)	The rate of canine retraction was significantly higher on the corticotomy side than on the CGr ($p < 0.05$).
Bajaj et al. (2022) [31]	split mouth RCT	30	18–25	MOPs and PBM	MBT prescription brackets	The retraction rate is 1.1 times higher with MOPs than with PBM.

Table 2. *Cont.*

Author (Year)	Study Design	Number of Patients	Average Age (Years)	Dental Movement Acceleration Techniques	Orthodontic Treatment	Outcomes
Bhad (Patil) e Karemore (2022) [32]	A clinical study with a split-mouth design	19	18–24	PEMF therapy	SW; NiTi closed-coil springs.	The rate of OTM in the experimental group was significantly higher than the CGr, with a mean increase in M1 of 41% and a mean increase in M2 of 31%.
Bhattacharya (2014) [33]	RCT	20	18.8 ± 3.48	Corticotomy	MBT prescription brackets; NiTi closed coil spring 250 g force; medium anchorage (transpalatal arch)	The corticotomy group's meantime for en masse retraction was 131 ± 7.5 d, compared to 234 ± 9 d for the traditional approach.
Chandran (2018) [34]	RCT	20	14.5	Bur corticotomy	MBT prescription brackets; active tie-back 100 g force	Alveolar corticotomy enhanced the rate of canine retraction by about 40%.
Cruz (2004) [14]	RCT	11	15	LLLT	Roth prescription brackets from right to left canines;12 mm NiTi closed coil spring	Laser Group is faster than CGr with a ratio of 1.34.
Farhadian et Al. (2021) [35]	RCT	60	LLLT group (20.9 ± 5.5); LED group (21.7 ± 4.2); CGr (22.7 ± 5.3).	LLLT Group: GaAlAs (810 nm; 100 mW) performed on days 0, 3, 30, and 60. LED Group: intraoral LED device (wavelength: 640 nm; 10 j/cm^2; 40 mW/cm^2), 5 min/day	MBT and Roth prescription brackets. Medium anchorage (trans-palatal arch, on second molars) 6-mm NiTi closed-coil spring 150 g force	The laser group had a considerably higher rate of canine retraction than the CGr (p = 0.004). This variable is also 26% higher in the LED group than in the CGr; the difference is not statistically significant (p = 0.17).
Farid et Al. (2019) [36]	RCT Split mouth	16	21.5 ± 3.2	LLLT: In-Ga-As diode laser (940 nm; 0.5 W/cm^2 power density, 5 J/cm^2 Fluence, CW, 240 s time irradiations), weekly for the first month and twice monthly for the next three months	Roth prescription brackets; Medium anchorage (trans-palatal arch). 6-mm NiTi closed-coil spring	LLLT paired with corticotomy did not achieve a higher rate of canine retraction than the gold standard corticotomy approach alone.
Feizbakhsh et al. (2018) [37]	RCT	20	28	MOPs	Roth prescription brackets	The retraction rate was twice as high in the MOPs group than in the control group.

Table 2. *Cont.*

Author (Year)	Study Design	Number of Patients	Average Age (Years)	Dental Movement Acceleration Techniques	Orthodontic Treatment	Outcomes
Gibreal (2019) [38]	RCT	34	16–27	Piezocision	MBT prescription brackets; power chain	59% less TT in piezocision group.
Gibreal (2022) [39]	Parallel-group RCT	34	20.86	3D-guided piezo-assisted orthodontic treatment/conventional orthodontic	MBT prescription brackets; 5 incisions in the labial cortical plate between the six anterior teeth.	OTM time was decreased by 48% in the experimental group. This could be explained via the regional acceleratory phenomenon (RAP) following the intentional bone injury.
Hasan et A. (2017) [40]	RCT	26	20.07 ± 3.13	LLLT: 830 nm; 2.25-J/cm^2	MBT prescription brackets	LLLT is an efficient way to accelerate OTM.
Impellizzeri et al. (2020) [41]	RCT	3	16	LLLT	SW brackets; lace-back	After 1 month of follow-up, the laser side was 32% faster than the placebo.
Isola et Al. (2019) [42]	RCT	41	13.4 ± 2.1	LLLT: 810 nm laser applied on 3 points (1 W, continuous wave 66.7 J/cm^2; 8 J) at 3, 7, and 14 days and every 15 days until the space closed.	Self-ligating brackets system; Closed NiTi coil spring (9 mm long, 50 N).	LLLT therapy is effective in accelerating OTM.
Khera et al. (2022) [43]	RCT	25	18–25	A customised vibratory device is similar to AcceleDent Aura, with a frequency of 30 Hz and force of 0.25 N (25 g).	0.018″ MBT prescription brackets	There is no statistically significant difference between the experimental and CGrs.
Kumar et al. (2020) [44]	RCT	65	Group 1 (17 ± 0.80), Group 2 (17.40 ± 0.7), Group 3 (16.90 ± 1.1)	Low-frequency vibrations (30 Hz) using a custom-made vibratory device	Group 1: Passive self-ligating brackets (MBT prescription) with low-frequency vibrations Group 2: Conventional MBT brackets with low-frequency vibrations Group 3: Conventional MBT brackets without low-frequency vibrations	There are no significant differences in the rate of space closure between the three groups ($p > 0.05$).
Kundi et al. (2020) [45]	Parallel group RCT	30	27.5 ± 4.4	MOPs	MBT prescription brackets	Acceleration of OTM by 2–3 times.

Table 2. *Cont.*

Author (Year)	Study Design	Number of Patients	Average Age (Years)	Dental Movement Acceleration Techniques	Orthodontic Treatment	Outcomes
Kurt et al. (2017) [46]	P	33	DAD group: 15.8 ± 1.96; DG group: 16.02 ± 2.8	The distraction of the Alveolar Bone (DAD) and Distalisation group (DG)	SW Brackets	DAD Group: Canines retracted 7.9 ± 1.49 mm in 11.8 ± 1.3 days; DG group: Canine distalisation achieved 5.29 ± 2.01 mm in 200 ± 57 days; significant distal displacement of maxillary incisors (1.96 ± 2.79 mm) and canines (5.29 ± 2.01 mm).
Le et Al. (2023) [47]	CS	16	22.53 ± 3.54	LLLT: GaAlAs laser (810 nm; 100 mW continuous mode, twice-a-month irradiation; 5.1 J/cm^2	MBT prescription bracket	In orthodontic therapy, LLLt had a positive influence on OTM speed.
Liao et al. (2017) [48]	CS	13	13.6	Vibration using an Oral B (USA) Hamming Bird Vibrating Unit	Coil springs attached to maxillary first molar and canine brackets	OTM was higher with vibration compared to non-vibration.
Mahmoudzadeh et al. (2020) [49]	RCT	12	18.91 ± 3.87	Laser corticotomy	MBT prescription brackets; 9-mm-long nickel-titanium closed coil springs	At one month, OTM under laser was 2.5 times higher than the control.
Moradinejad et Al. (2022) [50]	RCT	32	19.13 ± 2.27	LLLT + Piezocision	MBT prescription bracket; short-size elastic chain	Piezocision is superior in accelerating movement compared to LLLT. Speed is higher with the combination of piezocision and LLLT.
Qamruddin et Al. (2021) [51]	CS	20	20.25 ± 3.88	LLLT: GaAlAs (940 nm; 100 mW for 3 s)	MBT prescription brackets; NiTi closed-coil spring 150 g force	The use of LLLT at regular orthodontic sessions (3 weeks apart) speeds up OTM.
Qamruddin et Al. (2017) [52]	RCT	22	19.8 ± 3.1	LLLT: GaAlAs laser (940 nm) applied at baseline and then repeated after three weeks for two more consecutive follow-up visits	Self-ligating brackets; 6 mm NiTi closed coil springs 150 g force	LLLT applied at 3-week intervals can accelerate OTM.

Table 2. *Cont.*

Author (Year)	Study Design	Number of Patients	Average Age (Years)	Dental Movement Acceleration Techniques	Orthodontic Treatment	Outcomes
Naji et al. (2022) [53]	RCT	40	21.3 ± 1.8	Injection of PRF and PRP	Roth 0.018-inch brackets; Ricketts Retraction Spring (Blue-Elgiloy, 0.016 * × 0.022 inches)	PRP determined a more pronounced acceleration of canine retraction than i-PRP.
Sakthi et al. (2014) [54]	RCT	40	n.d.	Bur decortication	Roth prescription; NiTi closed coil spring 250 g force; no anchorage	The average space closure velocity in the maxilla was 1.8 mm/month, and the mandible was 1.57 mm/month, compared to 1.02 mm/month in the maxilla and 0.87 mm/month in the CGr.
Simre (2022) [55]	RCT	24	20.50 ± 2.58	Piezocision-conventional bur corticotomy	SW brackets; NiTi closed coil springs	Corticotomy with bur was 1.5–2 times more rapid, whereas piezocision was 1.5 times faster.
Storniolo-Souza (2020) [56]	RCT	11	14.04	LLLT	SW brackets; NiTi closed coil springs (12 mm length)	High retraction speed of the mandibular canine laser side.
Subrahmanya (2020) [57]	P	15	18–26	Piezoincision	SW brackets; elastomeric chain 150 g force; medium anchorage (BTP)	1.5 times acceleration of movement.
Sultana (2022) [58]	RCT	13	20.83 ± 2.32	Piezoincision	MBT prescription Brackets; NiTi closed coil spring 250 g force; medium anchorage (transpalatal arch)	The piezocision group completed the levelling and alignment phase faster than the CGr.
Taha et al. (2020) [59]	Single-center pilot RCT	21	15.09 ± 1.7 CGr and 15.9 ± 1.29 in the ExGr	AcceleDent Aura (OrthoAccel Technologies Inc., Bellaire, USA) is used in the ExGr for 20 min daily.	MBT prescription brackets	There were no statistically significant differences in OTM between the control and ExGrs: 1.21 ± 0.32 mm/month in the CGr and 1.12 ± 0.20 mm/month in the ExGr.
Uday H Barhate et al. (2022) [60]	RCT	15	18–25	Injection of L-PRF	Standard Edgewise appliance of 0.018″slot dimension	A slight acceleration was found in the first four weeks.
Varella (2018) [61]	P	10	17.7 years	LLLT	MBT prescription Brackets; 9-mm-long NiTi closed coil spring	The laser side is two times faster than the control side (C Side) with high production of IL-1b.

Table 2. *Cont.*

Author (Year)	Study Design	Number of Patients	Average Age (Years)	Dental Movement Acceleration Techniques	Orthodontic Treatment	Outcomes
Varughese et al. (2019) [62]	RCT	15	22.5	Periodontal injection of calcitriol (1.25 DHC) on the experimental side and injection of placebo gel on the C Side.	SW brackets; closed NiTi coil springs 150 g force	Significantly greater canine distalisation on the experimental side compared to the C Side.
Yassaei (2016) [63]	RCT	11	19 ± 4.21	LLLT	edgewise appliance; NiTi closed coil springs	LLT did not lead to statistically significant differences.
Zeitunlouian et al. (2021) [64]	RCT	21	20.85 ± 3.85	injection of PRF	MBT prescription Brackets	Statistically significant orthodontic movement acceleration at T2.

CGr: Control group; RCT: Randomised clinical trial; SW: Straightwire; MBT: McLaughlin–Bennett–Trevisi; NiTi: Nickel titanium; OTM: Orthodontic tooth movement; PRP: Platelet-rich plasma; LLLT: Low-level laser therapy; MOPs: Micro-osteoperforations; GaAlAs: Gallium aluminum arsenide; CT: Clinical trial; PEMF: Pulsed electromagnetic field; LED: Laser-emitting diode; TT: Treatment time; DAD: Distraction of the alveolar bone; DG: Distalisation group; CS: Case series; ExGr: Experimental group; PRF: Platelet-rich fibrinogen; P: Prospective study.

3. Results

The electronic database search generated 1672 results. Following duplication elimination, 1350 studies were screened for titles and abstracts. After the abstract screening, 1252 papers were rejected, and 82 articles were chosen for the eligibility evaluation. Following the full-text examination, 37 manuscripts were eliminated: 21 were off-topic, six had wrong settings, and 10 had no outcome of interest. Finally, 44 papers were chosen for the systematic review, divided for each acceleration technique into:

- Corticotomy, 15 articles
- PRF/PRP, 4 articles
- LLLT, 13 articles
- MOPs, 4 articles
- Vibration, 5 articles
- DAD, 1 article
- LLLT + corticotomy, 1 article
- Drugs, 1 article

Figure 2 summarises the selection procedure.

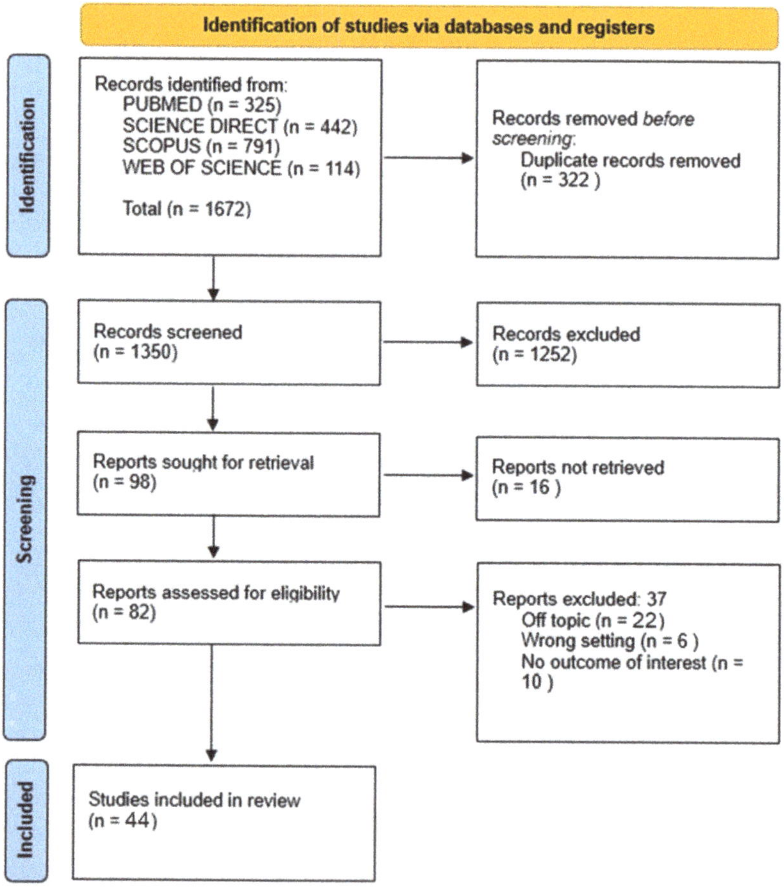

Figure 2. The literature search's Preferred Reporting Items for Systematic Reviews and Meta-Analyses (PRISMA) flow diagram.

4. Discussion

This systematic review examines different methods for accelerating OTM. In recent years, several techniques have been introduced that promise to speed up OTM, thereby reducing treatment duration. However, it is necessary to evaluate the effectiveness of these techniques via high-level randomised controlled studies to determine whether such devices can accelerate OTM.

4.1. Corticotomies

First introduced in 1892, corticotomy is a surgical technique involving a cortical bone cut, perforated or mechanically altered, with minimal involvement of the bone marrow [33]. It differs from osteotomy, where the cut involves bone and marrow. Regional Accleratory Phenomena (RAP) is based following corticotomy surgery on the principle that rapid bone remodelling and increased cell turnover occur following any trauma to bone tissue. There is a 4–5-month window during which the bone physiology changes, where the trabecular bone loses density and selectively offers less resistance to OTM. In AOO (Accelerated Osteogenic Orthodontics), non-activated teeth provide a relative anchorage for activated teeth that move faster (Figure 3) [33,65,66].

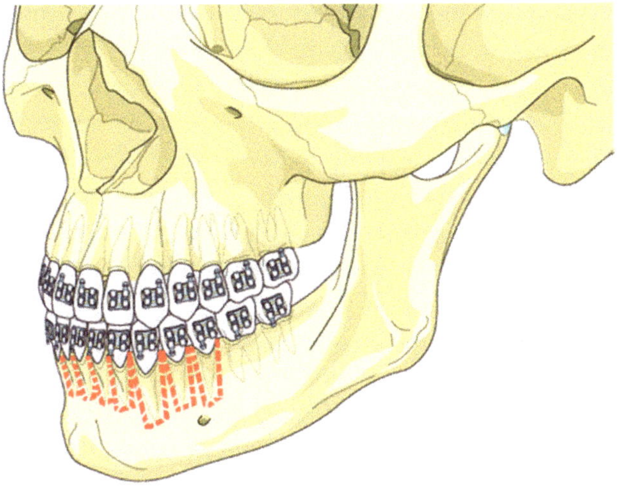

Figure 3. Corticotomies technique, as shows the red dashes.

Several studies have been conducted to investigate the efficacy of this strategy. Bhattacharya et al. (2014) evaluated the impact of corticotomy on a sample of 20 patients [33]. In the group of ten patients who underwent surgery, after lifting a full-thickness flap, a 2 mm round bur was used to make incisions between the inter radicular spaces from premolar to premolar, starting 2 mm apical from the bone crest and ending more than 2 mm from the root apex. The cuts for horizontal corticotomy involved both labial and palatal sides, and demineralised freeze-dried bone was applied before flap closure and suturing. Once compared to the CGr, the ExGr TT was significantly shorter(131 ± 7.5 days vs. 234 ± 9) [33]. Chandran et al. (2018) compared corticotomy versus the control in a split-mouth study [34]. Two vertical incisions were made, one on the canine distal line angle and the other on the second premolar mesial line angle, and full-thickness mucoperiosteal flaps were reflected. Selective decortication was performed on the buccal and lingual cortical plates using a fissure bur (#556) in a high-speed handpiece. To standardise all corticotomy operations, decortication was conducted at three sites: the buccal plate, the crest of the alveolar ridge, and the palatal plate (Figure 1). The study concludes that corticotomy accelerates OTM by 40% [34,67]. Sakthi et al. (2014), which analysed bur decortication via 701 slit burs and number 2 round burs attached on a micromotor handpiece, reached a similar conclusion, as it measures the average space closure velocity in the maxilla as 1.8 mm/month, and in the lower jaw, as 1.57 mm/month, compared to 1.02 mm/month in the maxilla and 0.87 mm/month in the CGr [54].

Addanki et al. (2017) inserted an additional variable in the split mouth RCT study, as it compared bur corticotomy performed only on the buccal side with corticotomy performed on both the buccal and palatal side and came to the conclusion that there are no statistically notable differences within the two populations [24].

Dibart pioneered piezocision as a minimally invasive approach for accelerating OTM. Piezocision is a potentially minimally invasive tooth acceleration procedure because of its numerous periodontal, cosmetic, and orthodontic benefits [57]. Incisions are made in the buccal mucosa under local anaesthetic 2–3 mm below the interproximal papilla's base. To decorticate the alveolar bone, the tip of the piezotome is placed at a depth of 3 mm. Subrahmanya's study (2020) states that piezocision accelerates canine retraction movement by 1.5 times [57].

In an RCT study, Simre et al. (2021) compared traditional corticotomy with bur using piezocision. Buccal 1 cm transmucosal incisions were made distal to the canine and mesial to the second premolar area in both groups [55]. The incisions were made 5 mm apical to the papillae, and the vertical guideline bur holes were drilled and joined on the buccal cortex using a No.8 tungsten carbide round bur mounted on a straight handpiece and rotated at 30,000 rpm. At a depth of 2 mm, the groove went through the cortical bone, barely entering the spongiosa. The piezo group employed the OT7 (Mectron®, Carasco, Italy) piezo tip to produce a vertical groove over the buccal cortex.

This study assessed that corticotomy with bur was 1.5–2 times more rapid, whereas piezocision was 1.5 times faster; both are effective therapeutic options [55].

The Gibreal (2019) RCT study compared piezocision with traditional orthodontics in the mandible in 36 patients [38]. Radiographic-guided micro piezoelectric corticotomies on the labial surfaces of the alveolar bone between the six anterior teeth were performed on patients in the ExGr to accelerate alignment. Compared to traditional therapy, this approach required 59% less TT to correct highly crowded lower anterior teeth [38].

Aksakalli (2015) compared two minimally invasive techniques: in a split-mouth study on ten patients, he performed traditional corticision with a number 15 blade in the mesiobuccal and distobuccal area of the maxillary canines on one side; on the contralateral side, he performed piezocisions of a depth of 3 mm [7]. The superimposition of the 3D models, carried out using the third palatine wrinkle as a reference point revealed approximately twice as fast movement on the piezocision side as on the C Side, especially during the first month of treatment [7].

Sultana et al. (2022) evaluated the effectiveness of piezocision compared to conventional orthodontics [58]. The piezocision group completed the levelling and alignment phase substantially faster than the CGr (mean difference = 31.5 days, 95% CI 6.5, 56.5; $p = 0.018$). The alignment rate in the piezocision group was quicker in the first two months than in the CGr [58].

The Abbas study (2015) compared piezocision and control versus corticotomy and control [23]. They assessed that orthodontics supported via corticotomies are 1.5 to 2 times faster than traditional orthodontics, and piezocision was 1.5 more rapid than conventional orthodontics [23].

Also, Al Imam (2019) compared piezocision with traditional orthodontics: the ExGr with piezocision significantly improved the rate of incisor retraction by 53%, while the retraction time was significantly reduced by 27% [25].

Gibreal et al. (2022) used a new technique, with 3D-guided piezo-assisted orthodontic treatment, compared to conventional orthodontic [39]. In ExGr 5, incisions were made in the labial cortical between the anterior teeth. The study revealed an OAT reduced 48% in this group compared to the CGr. Through a slight incision, this procedure might enable the creation of a simple, precise, predictable, and safe localised alveolar decortication. As a result, no harm would be inflicted on any surrounding anatomical structures if a flap was not raised [39].

The RCT study by Mahmoudzadeh (2020) that analyses the influence of lasercision on OTM is exciting. Lasercision corticotomy (Er, Cr: YSGG 3.5 W, 30 Hz, 40% air, 80% water) in one maxillary quadrant required 59% less TT to correct highly crowded lower anterior teeth than traditional therapy [49].

Alfawal (2018) analysed laser-assisted flapless corticotomy (LAFC) with ER: YAG laser and piezocision [26]. In both groups, the rate of canine retraction was two-fold higher in the ExGr than in the CGr in the first month and 1.5-fold higher in the second month ($p < 0.001$). In addition, the total canine retraction time was decreased by nearly 25% in both groups when comparing the experimental and C Sides [26].

Baeshen (2020) conducted a partial buccal plate corticotomy distal to the lingual vertical and subapical incisions, and the lingual flap was not raised [30]. This technique revealed that the rate of canine retraction on the corticotomy side was substantially greater than on the C Side [30].

The literature [26,38,68] confirms the efficacy of corticotomy, which reduces treatment time by 40 per cent; bur corticotomy is 1.5–2 times faster, and piezocision is 1.5 times faster, with both being effective. Piezocision accelerates canine retraction by 1.5 times. Lasercision requires 59% less TT for crowded anterior teeth. Corticotomy can be used in different clinical contexts and with different types of orthodontic treatment, providing a customised acceleration solution. However, it also has some disadvantages and risks associated with surgery. It is a technique that requires a learning curve on the part of the clinician, and it can still cause discomfort with oedema and pain for the patient.

4.2. Micro-Osteo-Perforations (MOPs)

Bajaj et al. (2022) compared the effects of photobiomodulation (PBM) and micro-osteo perforations (MOPs) on the speed of canine retraction in a study group of 30 patients [31]. The extraction of the premolars was followed by a waiting period of 3 months to allow trabecular bone formation. Three vertical MOPs (approximately 19-gauge diameter) were performed distal to the canine root using Propel contra-angle perforation screws in the MOPs group. In the PBM group, the canine was stimulated using a Gallium Aluminium Arsenide (GaAlAs) semiconductor diode laser for 10 s at ten different points of the canine root. The retraction rate was approximately 1.1-fold higher in patients treated with MOPs than in comparison patients [31].

Feizbakhsh et al. (2018) analysed accelerated canine distalisation using MOPs (Figure 4) [37]. After 28 days, the movement was assessed by analysing the digital models of the two arches using the canine and second premolar at three locations as retrievals. MOPs increased OTM over twice as much as the C Side [37].

In the study by Attri et al. (2018), the retraction was activated immediately after the MOPs were executed. MOPs were carried out using a manual screwdriver (Propel device) and drilling screws. Each patient received three perforations (1.5 mm in diameter and 2–3 mm deep) in the extraction space at an equal distance from the canine and second premolar at the level of the bony cortical performed every 28 days until the space was closed entirely. When MOP patients were compared to the control individuals, there was an increase in OTM. Therefore, the authors suggest its use after carefully evaluating the risk–benefits, as there is an increase in costs and initial discomfort during the procedure [29]. Kundi et al. (2020) compared the extent of canine distalisation in patients undergoing cortical perforations without flaps or MOPs.

Patients were divided into an intervention group and a CGr. In one session, three MOPs (diameter 1.5 mm, depth 2.5 mm) were executed distal to the canine on both sides using the Propel device at the buccal cortical level. The mesial movement of the molar associated with canine retraction was also investigated using the median palatine line and the most prominent part of the palatine wrinkles as references.

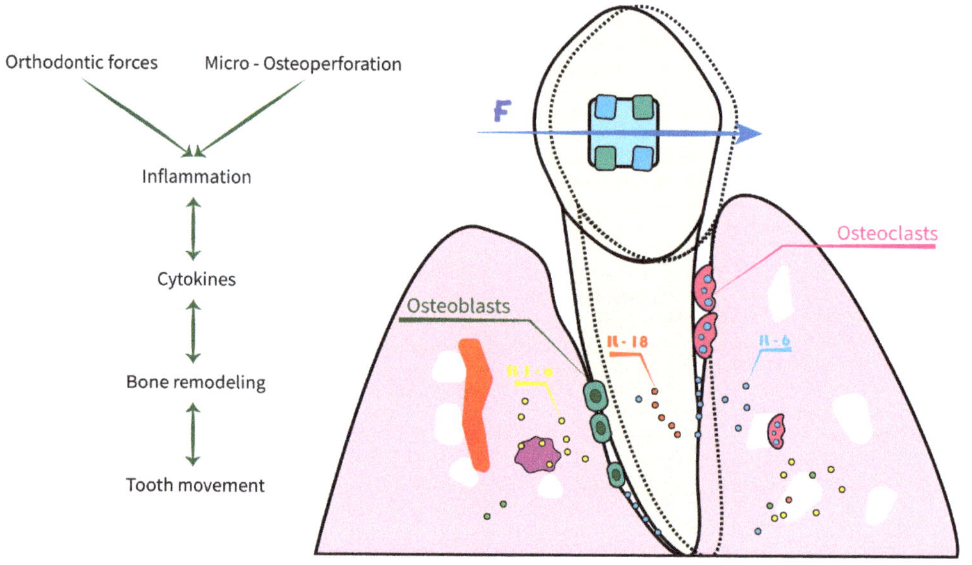

Figure 4. Mechanism of action of MOPs.

The orthodontic movement was accelerated 2–3 times in the MOPs group compared to the CGr [45]. Compared to the other articles used in this section of the discussion, Bajaj's work did not compare the improvement in the speed of canine retraction with a CGr treated only with fixed therapy. Still, it compared with a group exposed to PBM [31]. It can be deduced, however, that MOPs are a valid aid in increasing the speed of tooth retraction, although particular attention must be paid to reducing pain when performing this procedure.

MOPs have shown promise in accelerating orthodontic movement [8,69,70]. These are simple and minimally invasive procedures; however, they may cause temporary discomfort, swelling, or slight pain at the treated site.

4.3. Vibration Therapy

Several trials suggest that mechanical vibratory devices, in their current setting, do not offer significant advantages for orthodontic OTM. In the study by Khera et al. (2022), the effect of a customised vibratory device on the speed of OTM, particularly during canine retraction, was analysed [43]. The canine retraction was initiated at least four months after the first maxillary premolars extraction to eliminate any effects due to post-extraction regional acceleratory phenomena. Additionally, a customised vibratory device was used to reduce cost, which was economically advantageous compared to commercial devices and maintained a frequency of 30 Hz and a force of 0.25 N.

The study's primary objective was to evaluate the effect of low-frequency vibrations (30 Hz) on the speed of OTM through the customised vibratory device by comparing canine retraction between the vibrated and non-vibrated sides. The present study concluded that low-frequency vibratory stimulation (30 Hz) applied for 20 min per day using a customised vibratory device does not significantly accelerate the rate of canine retraction [43]. The trial conducted by Taha et al. (2020) aimed to compare two groups of adolescents undergoing complete orthodontic treatment with and without using the AcceleDent Aura device [59,71]. The study results showed that the total amount of OTM did not show statistically significant differences between the groups at any of the three time intervals. The study attempted to minimise bias and observed an average monthly OTM rate of 1.21 ± 0.32 mm in the CGr and 1.12 ± 0.20 mm in the ExGr.

In conclusion, using the AcceleDent Aura device did not significantly affect accelerating maxillary canine retraction or reducing perceived pain during orthodontic treatment [59]. Liao et al. (2017) investigated the effects of vibration-enhanced OTM and the underlying mechanisms [48]. The vibration was applied to the buccal surface of canines for 10 min daily for 28 days, using an Oral B (USA) Hamming Bird vibrating unit. The amplitude of the vibrating force was measured at approximately 0.2 N (20 g), and the frequency of the vibrations was 50 Hz.

The results showed that the total space closure and distalisation of canines were significantly more significant on the vibration side than on the non-vibration side. The research concluded that applying low to medium-frequency vibration, such as 50 Hz in this study, can accelerate OTM without causing adverse effects like tissue necrosis or other undesirable outcomes [72]. The study suggested that the mechanism for OTM acceleration may be more biologically based than mechanically based, as the short duration of vibration application seems to stimulate OTM-related cells and factors via temporarily sustained and dynamic amplification of the pressure levels within the periodontal ligament (PDL). Therefore, applying mechanical vibrations could increase the speed of OTM and may be considered a promising method for accelerating extraction-based orthodontic treatments [48]. Kumar et al. (2020) evaluated the rate of orthodontic movement in adolescent patients combined with low-frequency mechanical vibrations in passive self-ligating and conventional appliances [44].

The customised vibrating device used had a frequency of 30 Hz and was used by patients for 20 min per day during the space closure phase. The primary objective was to measure the space closure rate (mm/month). The results showed no statistically significant differences regarding the space closure rate among the three groups.

Therefore, the null hypothesis was confirmed, namely that there was no difference in the rate of orthodontic movement between passive self-ligating and conventional appliances in patients with low-frequency vibrations [44,73]. Another method described is the pulsed electromagnetic field (PEMF) used to reduce TT, as described in the study by Bhad (Patil) and Karemore (2022) [32].

An electrical engineer designed a device to generate a weak PEMF of 0.5 mT (Tesla), at 1 Hz. PEMF therapy was initiated on the same day as the application of the closed-coil spring, and the PEMF device consisted of an integrated circuit powered via a battery and embedded in a removable acrylic appliance. Patients were required to wear the appliance for 8 h at night, and the device was checked at each appointment. The results showed that the ExGr, exposed to the PEMF, experienced a 1.2-fold increase in the rate of OTM compared to the CGr, translating to a 41% increase.

The average time for canine retraction in the ExGr was 4.5 months, while the CGr took 6–6.5 months. The study concludes that PEMF therapy can physiologically increase the rate of OTM, thereby reducing overall TT. When used in combination with closed-coil springs, 1 Hz PEMFs were successful in increasing OTM. The study suggests that PEMF therapy could be safely and routinely used during orthodontic treatment to shorten TT [32].

El-Angbawi et al. [74] analysed two studies that compared the use of OrthoAccel and Tooth Masseuse devices with standard orthodontic mechanics during alignment and canine retraction stages, respectively. The trials evaluated tooth mobility objectively, but meta-analysis was hampered by varying outcome measures at various stages of treatment. Additionally assessed were discomfort, pain, and negative consequences. There needed to be a discussion of duration or how frequently to visit. Over ten weeks, the Tooth Masseuse improved lower incisor alignment with minimal pain variations. The maxillary canine movement was slightly faster with OrthoAccel, with no clinical significance.

In summary, although vibration therapy offers a non-invasive and user-friendly approach, its effectiveness for orthodontic tooth movement remains inconclusive [75–77]. Despite a customised device's economic advantage, low-frequency (30 Hz) vibratory stimulation for 20 min daily did not significantly accelerate canine retraction. Pulsed electromagnetic field (PEMF) therapy increased the OTM rate by 1.2-fold, shortening treatment time.

PEMF therapy, especially when combined with the other methods, offers the potential for accelerated orthodontic treatment.

4.4. Low-Level Laser Therapy (LLLT)

LLLT has emerged as one of the most promising new supportive treatment techniques in recent years since it is a non-intervention therapy that is easy to obtain and does not require expensive equipment [47]. LLLT has been demonstrated to increase angiogenesis by up-regulating chemical mediators such as the vascular endothelial growth factor, to facilitate osteoclast and osteoblast cell proliferation and differentiation, and to accelerate OTM [35]. Furthermore, low-level laser irradiation (LLLI) has been demonstrated to benefit analgesics in various clinical and therapeutic applications.

LLLI reduces pain perception by preventing the release of arachidonic acid, which lowers prostaglandin E2 levels. It also causes the production of an endogenous opioid neuropeptide (beta-endorphin), which has powerful analgesic properties [52]. Several studies have been conducted to evaluate the impact of LLLT on improving OTM. Farhadian et al. worked an RCCT with 60 patients divided into three groups: 20 treated with LLLT, 20 treated with LED biostimulation, and 20 in the CGr. The extraction site and buccal surface of the canine were exposed to light using an intraoral LED device called Biolight®, which is comparable to Ortho-Pulse® and has a wavelength of 640 nm, an energy density of 10 j/cm^2, and a power density of 40 mW/cm^2 [35].

The patients were instructed to utilise the device for the maxillary dental arch for 5 min daily at the commencement of canine retraction. A GaAlAs diode laser with a wavelength of 810 nm and a power of 100 mW was used to treat the LLLT group. According to Farhadian et al., LLLT looked to help accelerate OTM by 60%, while the LED could not significantly speed up the process. In addition, patient-centred outcomes showed that neither LLLT nor LED impacted how painful the procedure was felt by the patient [35].

On a sample of 22 patients, Quamruddin et al. examined the effect of LLLI delivered at 3-week intervals on OTM and pain related to OTM using self-ligating brackets [52]. According to Qamruddin et al., LLLT is a helpful technique that, if used at intervals of three weeks, can double the rate of OTM [52]. To assess the utilisation of non-invasive or minimally invasive techniques to expedite OTM, such as LLLT, Moradinejad et al. conducted a split-mouth RCT [50]. Three parallel intervention groups were randomly assigned to 64 quadrants in 32 patients: LLLT, LLLT with piezoincision, and CGr. A 940 nm laser with 8 J and 0.5 W of power was utilised for 16 s at six sites to accomplish LLLT. This was performed on the first day and then again after three and six weeks.

This study demonstrated that although LLLT statistically and significantly sped up canine retraction and slowed anchoring loss, its effects were, at best, mild or moderate [50,78]. Lam et al. evaluated the effectiveness of LLLT on 16 patients using a split-mouth RCT [47]. A GaAlAs diode laser with an output power of 100 mW and an 810 nm wavelength was used to treat the LLLT group for 10 s on both the buccal and lingual surfaces. LLLT in the conditions was used in our study. In orthodontic therapy, a GaAlAs diode laser with a twice-monthly radiation dose of 5.1 J/cm2 positively impacted OTM speed [47]. Isola et al. assessed the effects of LLLT after extracting the first upper premolars for orthodontic purposes using a split-mouth RCT [42].

A diode laser operating in continuous wave mode at an 810 nm wavelength treated the test side at three places on the buccal and palatal sides at baseline, at 3, 7, and 14 days, and then every 15 days until the space closed. Only orthodontic traction was used on the C Side to treat the opposing chosen canine. This study shows that applying LLLT therapy successfully quickens OTM and lowers OTM-related discomfort levels [42]. In another split-mouth RCT, Qamruddin et al. evaluated the effects of LLLT [51].

After removing the first bicuspid on day 21, each canine was retracted using a 6 mm close coil NiTi spring stretched to 150 gm of force. Immediately following the spring installation on the experimental side, LLLT irradiation was used. A continuous, uninterrupted beam of light at a wavelength of 940 nm from a GaAlAs diode laser was used. Qamruddin et al. claimed that LLLT during routine orthodontic appointments spaced three weeks apart accelerates OTM and considerably lessens pain [51,79]. The combined impact of corticotomy and LLLT on the speed of OTM was examined by Farid et al. [36].

The premolar extractions were performed on the same day a surgical corticotomy was conducted. The laser was applied at the beginning of the four-month study period or on the first day of full canine retraction, as well as one week, two weeks, three weeks, and every two weeks after that. According to Farid et al., corticotomy and LLLT alone could not increase the rate of canine retraction above that of the gold-standard corticotomy approach [36]. Arumughan et al. evaluate whether the LLLT can accelerate OTM during en-masse retraction [28]. The experimental side was subjected to biostimulation by using a GaAlAs diode laser with a wavelength of 810 nm. Each site received ten irradiations for 10 s, five on the palatal side and ten on the buccal side of the tooth.

With a three-week interval between appointments, the total energy density at each application was 10 J. It was found that the rate of extraction space closing can be accelerated via biostimulation using an 810 nm diode laser. As a result, it can accelerate tooth mobility during orthodontic treatment [28]. Hasan et al. examined if LLLT might hasten the migration of crowded maxillary incisors during orthodontic treatment [40]. Patients in the laser group obtained an LLLT dosage from a GaAlAs laser device with an energy of 2 J/point at an 830 nm wavelength just after the first archwire was inserted. Due to a statistically significant difference in total TT between the two groups, Hasan et al. asserted that LLLT is a valuable technique for quickening OTM [40].

In 2020, Impellizzeri et al. used a split-mouth RCT to assess the efficacy of photobiomodulation therapy in accelerating OTM [41]. Following the extraction of the first premolar, canine retraction movement was monitored. Linear measurements of the canine's anteroposterior position were obtained at the onset of treatment and one month later, and these values were contrasted with those from the non-irradiated side to determine the pace of orthodontic therapy. GaAlAs PBM was administered four times to the experimental segment. The results indicated a statistically significant disparity and a 32% acceleration in OTM attributable to the biostimulatory agent [41].

Cruz et al.'s work (2004) also utilised LLLT to assess the speed difference between the irradiated and C Sides of OTM [14]. The authors emphasised a notable acceleration of OTM on the side that had been exposed to radiation, and they specifically demonstrated that the laser group always experienced faster displacement with each spring activation. The authors claimed that LLLT in that region increased the region's receptivity to biochemical alterations that help OTM [14]. Strniolo-Souza et al. (2020) monthly applied LLLT in both the upper and lower arch [56].

The LLLT technique for the upper arch varied depending on whether the side was buccal or palatal; the latter side required a larger dosage since the canine root was further from the laser application site due to the greater bony thickness of the palate. However, they discovered in their research that the LLLT procedure proved only helpful in the upper arch in the first stages of canine retraction, with speed around the same as on the C Side. The effectiveness of LLLT is correlated with both the dosage and frequency of laser application, according to scientists, who speculated that variations in bone density between the maxilla and mandible may have had differing effects on laser light absorption [56].

When LLLT photonic radiation enters the cell nucleus, it increases the production of ribonucleic acid (RNA), deoxyribonucleic acid (DNA), and protein synthesis. Enhancing the inflammatory response to specific stimuli produces biostimulating effects on the cellular metabolic processes [41].

Inflammatory responses mediate orthodontic movement; some research has sought to identify the primary inflammatory players responsible for increased OTM following LLLT. With the use of the GaAlAs diode laser, Varella's (2018) study intended to measure and analyse the levels of IL-1β in gingival crevicular fluid during OTM [61]. In the split-mouth experiment, the canine retraction on the irradiation side progressed faster than on the C Side.

Crevicular gingival fluid samples were examined using ELISA assays, and it was discovered that the amount of IL-1β was more significant in the laser-irradiated region. The scientists hypothesised that the dental acceleration was likely caused by the increased IL-1β levels induced by LLLT [61]. In contrast, in a previous study, interleukin-6 was analysed as potentially responsible for the acceleration of OTM after LLLT [63].

Patients underwent maxillary canine retraction to close the extraction space, in which only one side had LLLT applied. During orthodontic treatment, gingival crevicular fluid samples were taken to analyse the IL-6 levels. The results showed that LLLT indeed accelerated OTM. However, the difference in the mean IL-6 concentration was not statistically significant, and it was impossible to attribute a major role to this cytokine [63].

LLLT is a non-invasive approach with potential benefits, including increased angiogenesis, osteoblast/osteoclast activity, and analgesic effects. Studies have explored LLLT's impact on Orthodontic Tooth Movement (OTM), indicating accelerated movement. However, challenges exist, including varying outcomes, precision demands, cost, availability, treatment duration, and frequency. Notably, LLLT's mechanism involves cellular responses and cytokine modulation, yet cytokine roles remain debated. While LLLT holds promise for enhancing OTM, carefully considering its advantages, disadvantages, patient preferences, and further research is essential for informed treatment decisions in orthodontic practice [80].

4.5. PRP and PRF

Leukocyte and platelet-rich fibrin (L-PRF) injections can accelerate OTM. The branchial vein collects 20 mL of blood, which is then centrifuged once at 700× g rpm for 3 min to obtain the L-PRF. To obtain PRF, which will subsequently be injected into the buccal and palate mucosa to quicken orthodontic movement, the yellow–orange section of the animal was employed [81]. Uday H. Barhate et al. (2022) evaluated the effect of L-PRF on the rate of canine maxillary retraction [60]. A careful analysis of the study models observed some correlation but is not statistically significant between canine retraction and the concentration of cytokines such as IL-1β and TNF-α.

Acceleration mainly occurred in the first four weeks; after that, retraction between the experimental and C sides was equal [60]. Zeitunlouian et al. (2021) performed a similar split-mouth experimental study to evaluate the effect of PRF on accelerating OTM during orthodontic treatment [64]. PRF was injected into the mucosa before canine retraction, and the procedure was repeated one month later. Results show a statistically significant acceleration occurred on the experimental side compared to the C Side [64].

On the other hand, Naji et al. (2022) compared the efficacy of PRF injection and PRP during orthodontic canine retraction [53]. A significantly faster rate of canine retraction was obtained with PRP infiltration in the first month. It was shown that PRP infiltration was more impactful with accelerated canine movement than PRF infiltration [53]. Angel et al. (2022) evaluated the effects of PRP on the acceleration of maxillary canine retraction [27,82]. The evaluation was performed by measuring the soluble receptor activator of nuclear factor κb (sRANKL) and the osteoprotegerin (OPG) ligand in the gingival crevicular fluid (GCF). After alignment, the premolar was extracted and then freshly prepared PRP was injected. The motion was assessed using digital model superposition at T0, T1 (30 days), and T3 (60 days). Movement occurred 35% more on the PRP than on the C Side, altering the OPG and sRANKL levels in GCF [27,83].

Therefore, the injection of PRF or PRP before a canine retraction in the post-extraction space may occur faster in the first few weeks. The prolonged PRF injections may be needed to achieve accelerated OTM, but this deserves more research.

Leukocyte and platelet-rich fibrin (L-PRF) injections show potential in hastening OTM initiation, with studies reporting accelerated movement [64,81,84]. Platelet-rich plasma (PRP) demonstrates initial superiority in canine retraction speed compared to PRF. Both interventions hold promise for modulating biomarkers associated with enhanced movement. Challenges encompass outcome variability, required treatment duration, and biomarker correlations. While early OTM acceleration with PRP and PRF is encouraging, a thorough evaluation considering patient preferences, long-term effects, and cost-effectiveness remains crucial. Ongoing research is vital for refining protocols and establishing sustained orthodontic benefits.

4.6. Drugs Therapy

Among strategies for OTM using drugs, the effect of vitamin D on canine distalisation and alveolar bone density using multi-slice spiral computed tomography (MSCT) has been studied. An RCT [62] used an in situ gel containing 1.25 DHC (active vitamin D metabolite) to evaluate its effect on OTM speed [85].

In the first month, the speed of canine movement was faster in the ExGr, but without a statistically significant difference compared to the CGr. In the following months, the rate of canine training in the ExGr was statistically significant. The effect of vitamin D was shown to be more significant when administered at doses close to normal physiological levels. In conclusion, the local administration of 1.25 DHC led to a significant increase in canine distalisation and a reduction in bone density in the trabecular bone tissue, suggesting that vitamin D may play a role in accelerating orthodontic movement [62].

4.7. Dentoalveolar Osteodistraction

The study by Kurt et al. (2017) aimed to evaluate the effects of the DAD device on OTM compared to conventional methods in patients with Class II malocclusion [46]. Thirty-three patients were divided into the DAD group and the Distalisation Group (DG). The patients who underwent a DAD have applied a retraction force of approximately 800 g for the upper canine. The rate of canine movement was measured in millimetres per month, and the dentoskeletal effects were assessed via an analysis of panoramic and lateral skull radiographs [86].

Furthermore, the dentoskeletal effects of DAD were limited to the treatment area, with no adverse effects on the position of the other teeth. The DAD group showed significantly faster canine movement (0.87 mm/month) concerning average speed experimented with conventional orthodontic technique (0.35/month) and no significant changes in the other maxillary and mandibular parameters except for maxillary canine retraction.

The DG group showed significant changes in the vertical dimension and mesial movement of the maxillary first molars, indicating anchorage loss [46,87]. No significant difference was found between the two groups in maxillomandibular measurement difference or root resorption. The study concludes that DAD can benefit patients with increased skeletal vertical dimensions and may reduce anterior tooth retraction time during fixed orthodontic therapy [46].

Osteodistraction, as exemplified via the DAD device, offers advantages in OTM for patients with Class II malocclusion [88]. The DAD group exhibited notably faster canine movement than conventional methods, benefiting those with increased skeletal vertical dimensions. Significantly, dentoskeletal effects were localised, avoiding negative impacts on the neighbouring teeth. However, some limitations were observed, including altered maxillary canine retraction and potential vertical dimension changes in the distalisation group. Root resorption and maxillomandibular measurements remained unaffected. While advantageous for specific cases, a careful consideration of possible anchorage loss and anatomical factors is essential.

Furthermore, this work [89] asserts that the absence of periodontal defects or endodontic lesions characterises the rapid distraction of the periodontal ligament during canine retraction. So, it is possible to rapidly distract the periodontal ligament without complications.

5. Limitation

This study reviews numerous RCTs and CSs regarding OTM acceleration techniques during extractive orthodontic treatments. However, orthodontic treatments that involved only medium and minimal anchorage appliances were considered in the literature selection, excluding orthodontic treatments that used orthodontic miniscrews as the maximum anchorage. Further studies will be needed to investigate this topic by considering this additional variable.

6. Conclusions

Most orthodontic therapies discussed in the literature seem promising and effective:

- Techniques like corticotomies and micro-osteo perforations (MOPs) exhibit 1.5 to 2 times faster acceleration than traditional methods.
- Piezoincisions are effective with variable success rates, offering time benefits but potential costs and discomfort.
- Vibrational therapy's impact on tooth movement is debated.
- Pulsed electromagnetic field significantly shortens treatment times.
- Low-level laser therapy speeds up tooth movement and offers analgesic benefits.
- PRF, PRP, and Vitamin D treatments increase movement speed.
- Dentoalveolar distraction aids shorter treatment, particularly in patients with vertical skeletal dimensions, minimizing anchoring loss.

These strategies expedite treatment, enhance patient experience, and minimise issues like cavities and enamel demineralisation. Research is needed to assess their efficacy, considering compliance, complications, risks, benefits, and efficacy in specific clinical contexts. Larger studies with standardised protocols can illuminate their impact on orthodontic care.

Author Contributions: Conceptualization, G.M.; I.F., F.V., A.M.C., A.D.I. and F.I.; methodology, A.P. (Assunta Patano), I.F., A.N., G.D. and A.M.I.; software, A.P. (Andrea Palermo), D.A., F.I. and G.D.; validation, G.M., S.C., A.D.I. and D.A.; formal analysis, G.M., A.N., D.A., A.M.C. and A.M.I.; investigation, A.D.I., A.P. (Assunta Patano), I.F., F.I. and G.D.; resources, A.D.I., F.V, I.F., A.N. and S.C.; data curation, I.F., A.N., D.A., A.M.C. and A.M.I.; writing—original draft preparation, F.V., A.P. (Assunta Patano), F.I. and A.M.I.; writing—review and editing, F.V., G.M., S.C., A.M.C. and G.D.; visualisation, A.D.I., A.N, S.C., D.A. and A.M.I.; supervision, I.F., F.V., A.N., A.P. (Andrea Palermo), S.C. and F.I.; project administration, G.M., A.P. (Assunta Patano), F.V., A.M.C. and G.D. All authors have read and agreed to the published version of the manuscript.

Funding: This research received no external funding.

Data Availability Statement: Not applicable.

Conflicts of Interest: The authors declare no conflict of interest.

Abbreviations

CGr	Control Group
C Side	Control Side
CT	Clinical trial
DAD	Dentoalveolar Distraction
DG	Distalisation group
DNA	Deoxyribonucleic acid
ExGr	Experimental Group
GaAlAs	Gallium Aluminum Arsenide
GCF	Gingival Crevicular Fluid
IL-1β:	L'interleuchina-1 beta
LAFC	Laser-assisted flapless corticotomy
LLLI	Low-level Laser Irradiation
LLLT	Low-level Laser Therapy
L-PRF:	Leukocyte and Platelet-Rich Fibrin
MOPs	Micro-osteoperforations
NiTi	Nickel-Titanium
NSAIDs	Nonsteroidal anti-inflammatory drugs
OPG	Osteoprotegerin
OTM	Orthodontic tooth movement

P	Prospective study
PBM	Photobiomodulation
PEMF	Pulsed Electromagnetic Field
PRF	Platelet-rich fibrinogen
PRP	Platelet-rich plasma
RAP	Regional Accleratory Phenomena
RCT	Randomised clinical trial
RNA	Ribonucleic acid
sRANKL	Soluble Receptor Activator of Nuclear factor κb
SW	Straightwire
TNF-α:	Tumor Necrosis Factor Alfa
TT	Treatment Time
VT	Vibration therapy

References

1. Wilcko, W.M.; Wilcko, T.; Bouquot, J.E.; Ferguson, D.J. Rapid Orthodontics with Alveolar Reshaping: Two Case Reports of Decrowding. *Int. J. Periodontics Restor. Dent.* **2001**, *21*, 9–19.
2. Coloccia, G.; Inchingolo, A.D.; Inchingolo, A.M.; Malcangi, G.; Montenegro, V.; Patano, A.; Marinelli, G.; Laudadio, C.; Limongelli, L.; Di Venere, D.; et al. Effectiveness of Dental and Maxillary Transverse Changes in Tooth-Borne, Bone-Borne, and Hybrid Palatal Expansion through Cone-Beam Tomography: A Systematic Review of the Literature. *Medicina* **2021**, *57*, 288. [CrossRef] [PubMed]
3. Mummolo, S.; Marchetti, E.; Albani, F.; Campanella, V.; Pugliese, F.; Di Martino, S.; Tecco, S.; Marzo, G. Comparison between Rapid and Slow Palatal Expansion: Evaluation of Selected Periodontal Indices. *Head. Face Med.* **2014**, *10*, 30. [CrossRef] [PubMed]
4. Fischer, T.J. Orthodontic Treatment Acceleration with Corticotomy-Assisted Exposure of Palatally Impacted Canines. *Angle Orthod.* **2007**, *77*, 417–420. [CrossRef]
5. Teixeira, C.C.; Khoo, E.; Tran, J.; Chartres, I.; Liu, Y.; Thant, L.M.; Khabensky, I.; Gart, L.P.; Cisneros, G.; Alikhani, M. Cytokine Expression and Accelerated Tooth Movement. *J. Dent. Res.* **2010**, *89*, 1135–1141. [CrossRef] [PubMed]
6. Miles, P.; Fisher, E. Assessment of the Changes in Arch Perimeter and Irregularity in the Mandibular Arch during Initial Alignment with the AcceleDent Aura Appliance vs No Appliance in Adolescents: A Single-Blind Randomized Clinical Trial. *Am. J. Orthod. Dentofac. Orthop.* **2016**, *150*, 928–936. [CrossRef] [PubMed]
7. Aksakalli, S.; Calik, B.; Kara, B.; Ezirganli, S. Accelerated Tooth Movement with Piezocision and Its Periodontal-Transversal Effects in Patients with Class II Malocclusion. *Angle Orthod.* **2015**, *86*, 59–65. [CrossRef]
8. Alikhani, M.; Raptis, M.; Zoldan, B.; Sangsuwon, C.; Lee, Y.B.; Alyami, B.; Corpodian, C.; Barrera, L.M.; Alansari, S.; Khoo, E.; et al. Effect of Micro-Osteoperforations on the Rate of Tooth Movement. *Am. J. Orthod. Dentofac. Orthop.* **2013**, *144*, 639–648. [CrossRef]
9. Quinzi, V.; Tecco, S.; Nota, A.; Caggiati, E.; Mummolo, S.; Marzo, G. Mesial Rotation of the Upper First Molar: Association with Anterior Dental Crowding in Mixed and Permanent Dentition. *Appl. Sci.* **2020**, *10*, 5301. [CrossRef]
10. Quinzi, V.; Ferro, R.; Rizzo, F.A.; Marranzini, E.M.; Federici Canova, F.; Mummolo, S.; Mattei, A.; Marzo, G. The Two by Four Appliance: A Nationwide Cross-Sectional Survey. *Eur. J. Paediatr. Dent.* **2018**, *19*, 145–150. [CrossRef] [PubMed]
11. Davidovitch, Z. Tooth Movement. *Crit. Rev. Oral Biol. Med.* **1991**, *2*, 411–450. [CrossRef] [PubMed]
12. Kau, C.; Nguyen, J.T.; English, J. The Clinical Evaluation of a Novel Cyclical Force Generating Device in Orthodontics. *Orthod. Pract. US* **2010**, *1.1*, 1–4.
13. Scarano, A.; Lorusso, F.; Inchingolo, F.; Postiglione, F.; Petrini, M. The Effects of Erbium-Doped Yttrium Aluminum Garnet Laser (Er: YAG) Irradiation on Sandblasted and Acid-Etched (SLA) Titanium, an In Vitro Study. *Materials* **2020**, *13*, 4174. [CrossRef]
14. Cruz, D.R.; Kohara, E.K.; Ribeiro, M.S.; Wetter, N.U. Effects of Low-Intensity Laser Therapy on the Orthodontic Movement Velocity of Human Teeth: A Preliminary Study. *Lasers Surg. Med.* **2004**, *35*, 117–120. [CrossRef]
15. Dohan Ehrenfest, D.M.; Rasmusson, L.; Albrektsson, T. Classification of Platelet Concentrates: From Pure Platelet-Rich Plasma (P-PRP) to Leucocyte- and Platelet-Rich Fibrin (L-PRF). *Trends Biotechnol.* **2009**, *27*, 158–167. [CrossRef]
16. Alissa, R.; Esposito, M.; Horner, K.; Oliver, R. The Influence of Platelet-Rich Plasma on the Healing of Extraction Sockets: An Explorative Randomised Clinical Trial. *Eur. J. Oral. Implantol.* **2010**, *3*, 121–134. [PubMed]
17. Kim, S.-J.; Park, Y.-G.; Kang, S.-G. Effects of Corticision on Paradental Remodeling in Orthodontic Tooth Movement. *Angle Orthod.* **2009**, *79*, 284–291. [CrossRef]
18. Wang, X.; Mei, M.; Han, G.; Luan, Q.; Zhou, Y. Effectiveness of Modified Periodontally Accelerated Osteogenic Orthodontics in Skeletal Class II Malocclusion Treated by a Camouflage Approach. *Am. J. Transl. Res.* **2022**, *14*, 979–989. [PubMed]
19. Boyne, P.J. Experimental Evaluation of the Osteogenic Potential of Bone Graft Materials. *Annu. Meet. Am. Inst. Oral. Biol.* **1969**, 13–21.
20. Quinzi, V.; Panetta, G.; Filippi, P.; Rizzo, F.A.; Mancini, L.; Mummolo, S. Autotransplatation of Immature Third Molars as Substitutes for Congenitally Missing Second Premolars: An Alternative Solution in a Young Patient with Oligodontia. *J. Biol. Regul. Homeost. Agents* **2020**, *34*, 155–163. [PubMed]

21. Liou, E.J.-W.; Figueroa, A.A.; Polley, J.W. Rapid Orthodontic Tooth Movement into Newly Distracted Bone after Mandibular Distraction Osteogenesis in a Canine Model. *Am. J. Orthod. Dentofac. Orthop.* **2000**, *117*, 391–398. [CrossRef] [PubMed]
22. Liberati, A.; Altman, D.G.; Tetzlaff, J.; Mulrow, C.; Gøtzsche, P.C.; Ioannidis, J.P.A.; Clarke, M.; Devereaux, P.J.; Kleijnen, J.; Moher, D. The PRISMA Statement for Reporting Systematic Reviews and Meta-Analyses of Studies That Evaluate Healthcare Interventions: Explanation and Elaboration. *BMJ* **2009**, *339*, b2700. [CrossRef] [PubMed]
23. Abbas, N.H.; Sabet, N.E.; Hassan, I.T. Evaluation of Corticotomy-Facilitated Orthodontics and Piezocision in Rapid Canine Retraction. *Am. J. Orthod. Dentofac. Orthop.* **2016**, *149*, 473–480. [CrossRef] [PubMed]
24. Addanki, P.; Gooty, J.R.; Palaparthy, R. Clinical and Radiographic Comparative Evaluation of Buccal and Palatal Corticotomy with Buccal Corticotomy in Periodontally Accelerated Osteogenic Orthodontics with Surgical Bur. *Contemp. Clin. Dent.* **2017**, *8*, 321–326. [CrossRef]
25. Al-Imam, G.M.F.; Ajaj, M.A.; Hajeer, M.Y.; Al-Mdalal, Y.; Almashaal, E. Evaluation of the Effectiveness of Piezocision-Assisted Flapless Corticotomy in the Retraction of Four Upper Incisors: A Randomized Controlled Clinical Trial. *Dent. Med. Probl.* **2019**, *56*, 385–394. [CrossRef] [PubMed]
26. Alfawal, A.M.H.; Hajeer, M.Y.; Ajaj, M.A.; Hamadah, O.; Brad, B. Evaluation of Piezocision and Laser-Assisted Flapless Corticotomy in the Acceleration of Canine Retraction: A Randomized Controlled Trial. *Head. Face Med.* **2018**, *14*, 4. [CrossRef]
27. Angel, S.L.; Samrit, V.D.; Kharbanda, O.P.; Duggal, R.; Kumar, V.; Chauhan, S.S.; Coshic, P. Effects of Submucosally Administered Platelet-Rich Plasma on the Rate of Tooth Movement. *Angle Orthod.* **2022**, *92*, 73–79. [CrossRef]
28. Arumughan, S.; Somaiah, S.; Muddaiah, S.; Shetty, B.; Reddy, G.; Roopa, S. A Comparison of the Rate of Retraction with Low-Level Laser Therapy and Conventional Retraction Technique. *Contemp. Clin. Dent.* **2018**, *9*, 260–266. [CrossRef]
29. Attri, S.; Mittal, R.; Batra, P.; Sonar, S.; Sharma, K.; Raghavan, S.; Rai, K.S. Comparison of Rate of Tooth Movement and Pain Perception during Accelerated Tooth Movement Associated with Conventional Fixed Appliances with Micro-Osteoperforations—A Randomised Controlled Trial. *J. Orthod.* **2018**, *45*, 225–233. [CrossRef]
30. Baeshen, H.A. The Effect of Partial Corticotomy on the Rate of Maxillary Canine Retraction: Clinical and Radiographic Study. *Molecules* **2020**, *25*, 4837. [CrossRef]
31. Bajaj, I.; Garg, A.; Gupta, D. Comparative Effect of Micro-Osteoperforation and Photo-Bio-Modulation on the Rate of Maxillary Canine Retraction: A Split Mouth Randomized Clinical Trial. *La Clin. Ter.* **2022**, *173*, 39–45. [CrossRef]
32. Bhad (Patil), W.A.; Karemore, A.A. Efficacy of Pulsed Electromagnetic Field in Reducing Treatment Time: A Clinical Investigation. *Am. J. Orthod. Dentofac. Orthop.* **2022**, *161*, 652–658. [CrossRef] [PubMed]
33. Bhattacharya, P.; Bhattacharya, H.; Anjum, A.; Bhandari, R.; Agarwal, D.K.; Gupta, A.; Ansar, J. Assessment of Corticotomy Facilitated Tooth Movement and Changes in Alveolar Bone Thickness—A CT Scan Study. *J. Clin. Diagn. Res.* **2014**, *8*, ZC26–ZC30. [CrossRef]
34. Chandran, M.; Muddaiah, S.; Nair, S.; Shetty, B.; Somaiah, S.; Reddy, G.; Abraham, B. Clinical and Molecular-Level Comparison between Conventional and Corticotomy-Assisted Canine Retraction Techniques. *J. World Fed. Orthod.* **2018**, *7*, 128–133. [CrossRef]
35. Farhadian, N.; Miresmaeili, A.; Borjali, M.; Salehisaheb, H.; Farhadian, M.; Rezaei-Soufi, L.; Alijani, S.; Soheilifar, S.; Farhadifard, H. The Effect of Intra-Oral LED Device and Low-Level Laser Therapy on Orthodontic Tooth Movement in Young Adults: A Randomized Controlled Trial. *Int. Orthod.* **2021**, *19*, 612–621. [CrossRef]
36. Farid, K.A.; Eid, A.A.; Kaddah, M.A.; Elsharaby, F.A. The Effect of Combined Corticotomy and Low Level Laser Therapy on the Rate of Orthodontic Tooth Movement: Split Mouth Randomized Clinical Trial. *Laser Ther.* **2019**, *28*, 275–283. [CrossRef]
37. Feizbakhsh, M.; Zandian, D.; Heidarpour, M.; Farhad, S.Z.; Fallahi, H.R. The Use of Micro-Osteoperforation Concept for Accelerating Differential Tooth Movement. *J. World Fed. Orthod.* **2018**, *7*, 56–60. [CrossRef]
38. Gibreal, O.; Hajeer, M.Y.; Brad, B. Efficacy of Piezocision-Based Flapless Corticotomy in the Orthodontic Correction of Severely Crowded Lower Anterior Teeth: A Randomized Controlled Trial. *Eur. J. Orthod.* **2019**, *41*, 188–195. [CrossRef]
39. Gibreal, O.; Al-modallal, Y.; Al-assaf, M. Evaluation of the Efficacy of 3D-Guided Piezosurgery in Accelerating Mandibular Orthodontic Teeth Alignment: A Randomized Controlled Trial in Adults. *Dentistry 3000* **2022**, *10*, 281–288. [CrossRef]
40. AlSayed Hasan, M.M.A.; Sultan, K.; Hamadah, O. Low-Level Laser Therapy Effectiveness in Accelerating Orthodontic Tooth Movement: A Randomized Controlled Clinical Trial. *Angle Orthod.* **2017**, *87*, 499–504. [CrossRef] [PubMed]
41. Impellizzeri, A.; Horodynski, M.; Fusco, R.; Palaia, G.; Polimeni, A.; Romeo, U.; Barbato, E.; Galluccio, G. Photobiomodulation Therapy on Orthodontic Movement: Analysis of Preliminary Studies with a New Protocol. *Int. J. Environ. Res. Public. Health* **2020**, *17*, 3547. [CrossRef] [PubMed]
42. Isola, G.; Matarese, M.; Briguglio, F.; Grassia, V.; Picciolo, G.; Fiorillo, L.; Matarese, G. Effectiveness of Low-Level Laser Therapy during Tooth Movement: A Randomized Clinical Trial. *Materials* **2019**, *12*, 2187. [CrossRef] [PubMed]
43. Khera, A.; Raghav, P.; Mehra, V.; Wadhawan, A.; Gupta, N.; Phull, T. Effect of Customized Vibratory Device on Orthodontic Tooth Movement: A Prospective Randomized Control Trial. *J. Orthod. Sci.* **2022**, *11*, 18. [CrossRef]
44. Kumar, V.; Batra, P.; Sharma, K.; Raghavan, S.; Srivastava, A. Comparative Assessment of the Rate of Orthodontic Tooth Movement in Adolescent Patients Undergoing Treatment by First Bicuspid Extraction and En Mass Retraction, Associated with Low-Frequency Mechanical Vibrations in Passive Self-Ligating and Conventional Brackets: A Randomized Controlled Trial. *Int. Orthod.* **2020**, *18*, 696–705. [CrossRef] [PubMed]
45. Kundi, I.; Alam, M.K.; Shaheed, S. Micro-Osteo Perforation Effects as an Intervention on Canine Retraction. *Saudi Dent. J.* **2020**, *32*, 15–20. [CrossRef]

46. Kurt, G.; İşeri, H.; Kişnişçi, R.; Özkaynak, Ö. Rate of Tooth Movement and Dentoskeletal Effects of Rapid Canine Retraction by Dentoalveolar Distraction Osteogenesis: A Prospective Study. *Am. J. Orthod. Dentofac. Orthop.* **2017**, *152*, 204–213. [CrossRef]
47. Le, L.; Le, K.; Do, T. Influence of Low-Level Laser Treatment on Tooth Movement in Orthodontic Treatment. *J. Int. Dent. Med. Res.* **2023**, *15*, 1614–1619.
48. Liao, Z.; Elekdag-Turk, S.; Turk, T.; Grove, J.; Dalci, O.; Chen, J.; Zheng, K.; Ali Darendeliler, M.; Swain, M.; Li, Q. Computational and Clinical Investigation on the Role of Mechanical Vibration on Orthodontic Tooth Movement. *J. Biomech.* **2017**, *60*, 57–64. [CrossRef]
49. Mahmoudzadeh, M.; Poormoradi, B.; Alijani, S.; Farhadian, M.; Kazemisaleh, A. Efficacy of Er,Cr Laser Incision Corticotomy in Rapid Maxillary Canine Retraction: A Split-Mouth Randomized Clinical Trial. *J. Lasers Med. Sci.* **2020**, *11*, 442–449. [CrossRef]
50. Moradinejad, M.; Chaharmahali, R.; Shamohammadi, M.; Mir, M.; Rakhshan, V. Low-Level Laser Therapy, Piezocision, or Their Combination vs. Conventional Treatment for Orthodontic Tooth Movement: A Hierarchical 6-Arm Split-Mouth Randomized Clinical Trial. *J. Orofac. Orthop.* **2022**. [CrossRef]
51. Qamruddin, I.; Alam, M.K.; Mahroof, V.; Fida, M.; Khamis, M.F.; Husein, A. Photobiostimulatory Effect of a Single Dose of Low-Level Laser on Orthodontic Tooth Movement and Pain. *Pain. Res. Manag.* **2021**, *2021*, 6690542. [CrossRef] [PubMed]
52. Qamruddin, I.; Alam, M.K.; Mahroof, V.; Fida, M.; Khamis, M.F.; Husein, A. Effects of Low-Level Laser Irradiation on the Rate of Orthodontic Tooth Movement and Associated Pain with Self-Ligating Brackets. *Am. J. Orthod. Dentofac. Orthop.* **2017**, *152*, 622–630. [CrossRef] [PubMed]
53. Naji, R.; Zeitounlouian, T.; Alomari, E.; Youssef, M. Evaluation of the Efficacy of Platelet-Rich Plasma (PRP) and Injectable Platelet-Rich Fibrin (i-PRF) in the Acceleration of Canine Retraction: A Randomized Controlled Trial. *J. Int. Oral. Health* **2022**, *14*, 243. [CrossRef]
54. Inchingolo, F.; Ballini, A.; Cagiano, R.; Inchingolo, A.D.; Serafini, M.; De Benedittis, M.; Cortelazzi, R.; Tatullo, M.; Marrelli, M.; Inchingolo, A.M.; et al. Immediately Loaded Dental Implants Bioactivated with Platelet-Rich Plasma (PRP) Placed in Maxillary and Mandibular Region. *Clin Ter* **2015**, *166*, e146-52.
55. Simre, S.S.; Rajanikanth, K.; Bhola, N.; Jadhav, A.; Patil, C.; Mishra, A. Comparative Assessment of Corticotomy Facilitated Rapid Canine Retraction Using Piezo versus Bur: A Randomized Clinical Study. *J. Oral. Biol. Craniofac Res.* **2022**, *12*, 182–186. [CrossRef] [PubMed]
56. Storniolo-Souza, J.; Lima, L.M.; Pinzan, A.; Alvarez, F.; Pereira, S.C.d.C.; Janson, G. Influence of Low-Level Laser Irradiation on Orthodontic Movement and Pain Level—A Randomized Clinical Trial. *Orthod. Waves* **2020**, *79*, 105–112. [CrossRef]
57. Inchingolo, F.; Tatullo, M.; Abenavoli, F.M.; Marrelli, M.; Inchingolo, A.D.; Gentile, M.; Inchingolo, A.M.; Dipalma, G. Non-Syndromic Multiple Supernumerary Teeth in a Family Unit with a Normal Karyotype: Case Report. *Int. J. Med. Sci.* **2010**, *7*, 378–384. [CrossRef]
58. Sultana, S.; Ab Rahman, N.; Zainuddin, S.L.A.; Ahmad, B. Effect of Piezocision Procedure in Levelling and Alignment Stage of Fixed Orthodontic Treatment: A Randomized Clinical Trial. *Sci. Rep.* **2022**, *12*, 6230. [CrossRef]
59. Taha, K.; Conley, R.S.; Arany, P.; Warunek, S.; Al-Jewair, T. Effects of Mechanical Vibrations on Maxillary Canine Retraction and Perceived Pain: A Pilot, Single-Center, Randomized-Controlled Clinical Trial. *Odontology* **2020**, *108*, 321–330. [CrossRef]
60. Barhate, U.H.; Duggal, I.; Mangaraj, M.; Sharan, J.; Duggal, R.; Jena, A.K. Effects of Autologous Leukocyte-Platelet Rich Fibrin (L-PRF) on the Rate of Maxillary Canine Retraction and Various Biomarkers in Gingival Crevicular Fluid (GCF): A Split Mouth Randomized Controlled Trial. *Int. Orthod.* **2022**, *20*, 100681. [CrossRef]
61. Varella, A.M.; Revankar, A.V.; Patil, A.K. Low-Level Laser Therapy Increases Interleukin-1β in Gingival Crevicular Fluid and Enhances the Rate of Orthodontic Tooth Movement. *Am. J. Orthod. Dentofac. Orthop.* **2018**, *154*, 535–544.e5. [CrossRef]
62. Varughese, S.T.; Shamanna, P.U.; Goyal, N.; Thomas, B.S.; Lakshmanan, L.; Pulikkottil, V.J.; Ahmed, M.G. Effect of Vitamin D on Canine Distalization and Alveolar Bone Density Using Multi-Slice Spiral CT: A Randomized Controlled Trial. *J. Contemp. Dent. Pract.* **2019**, *20*, 1430–1435. [PubMed]
63. Minervini, G.; Franco, R.; Marrapodi, M.M.; Fiorillo, L.; Cervino, G.; Cicciù, M. Economic Inequalities and Temporomandibular Disorders: A Systematic Review with Meta-analysis. *J. Oral. Rehabil.* **2023**, *50*, 715–723. [CrossRef] [PubMed]
64. Minervini, G.; Franco, R.; Marrapodi, M.M.; Crimi, S.; Badnjević, A.; Cervino, G.; Bianchi, A.; Cicciù, M. Correlation between Temporomandibular Disorders (TMD) and Posture Evaluated Trough the Diagnostic Criteria for Temporomandibular Disorders (DC/TMD): A Systematic Review with Meta-Analysis. *J. Clin. Med.* **2023**, *12*, 2652. [CrossRef] [PubMed]
65. Quinzi, V.; Saccomanno, S.; Manenti, R.J.; Giancaspro, S.; Coceani Paskay, L.; Marzo, G. Efficacy of Rapid Maxillary Expansion with or without Previous Adenotonsillectomy for Pediatric Obstructive Sleep Apnea Syndrome Based on Polysomnographic Data: A Systematic Review and Meta-Analysis. *Appl. Sci.* **2020**, *10*, 6485. [CrossRef]
66. Inchingolo, A.D.; Ferrara, I.; Viapiano, F.; Netti, A.; Campanelli, M.; Buongiorno, S.; Latini, G.; Carpentiere, V.; Ciocia, A.M.; Ceci, S.; et al. Rapid Maxillary Expansion on the Adolescent Patient: Systematic Review and Case Report. *Children* **2022**, *9*, 1046. [CrossRef]
67. Dinoi, M.; Marchetti, E.; Garagiola, U.; Caruso, S.; Mummolo, S.; Marzo, G. Orthodontic Treatment of an Unerupted Mandibular Canine Tooth in a Patient with Mixed Dentition: A Case Report. *J. Med. Case Rep.* **2016**, *10*, 170. [CrossRef]
68. Bakr, A.R.; Nadim, M.A.; Sedky, Y.W.; El Kady, A.A. Effects of Flapless Laser Corticotomy in Upper and Lower Canine Retraction: A Split-Mouth, Randomized Controlled Trial. *Cureus* **2023**, *15*, e37191. [CrossRef] [PubMed]

69. Alkebsi, A.; Al-Maaitah, E.; Al-Shorman, H.; Abu Alhaija, E. Three-Dimensional Assessment of the Effect of Micro-Osteoperforations on the Rate of Tooth Movement during Canine Retraction in Adults with Class II Malocclusion: A Randomized Controlled Clinical Trial. *Am. J. Orthod. Dentofac. Orthop.* **2018**, *153*, 771–785. [CrossRef] [PubMed]
70. Li, J.; Papadopoulou, A.K.; Gandedkar, N.; Dalci, K.; Darendeliler, M.A.; Dalci, O. The Effect of Micro-Osteoperforations on Orthodontic Space Closure Investigated over 12 Weeks: A Split-Mouth, Randomized Controlled Clinical Trial. *Eur. J. Orthod.* **2022**, *44*, 427–435. [CrossRef]
71. Pasini, M.; Giuca, M.R.; Ligori, S.; Mummolo, S.; Fiasca, F.; Marzo, G.; Quinzi, V. Association between Anatomical Variations and Maxillary Canine Impaction: A Retrospective Study in Orthodontics. *Appl. Sci.* **2020**, *10*, 5638. [CrossRef]
72. Campanella, V.; Gallusi, G.; Nardi, R.; Mea, A.; Di Taranto, V.; Montemurro, E.; Marzo, G.; Libonati, A. Dentinal Substrate Variability and Bonding Effectiveness: SEM Investigation. *J. Biol. Regul. Homeost. Agents* **2020**, *34*, 49–54. [PubMed]
73. Mummolo, S.; Nota, A.; Marchetti, E.; Padricelli, G.; Marzo, G. The 3D Tele Motion Tracking for the Orthodontic Facial Analysis. *Biomed. Res. Int.* **2016**, *2016*, 4932136. [CrossRef] [PubMed]
74. El-Angbawi, A.; McIntyre, G.T.; Fleming, P.S.; Bearn, D.R. Non-Surgical Adjunctive Interventions for Accelerating Tooth Movement in Patients Undergoing Fixed Orthodontic Treatment. *Cochrane Database Syst. Rev.* **2015**, *2016*, CD010887. [CrossRef] [PubMed]
75. Jing, D.; Xiao, J.; Li, X.; Li, Y.; Zhao, Z. The Effectiveness of Vibrational Stimulus to Accelerate Orthodontic Tooth Movement: A Systematic Review. *BMC Oral. Health* **2017**, *17*, 143. [CrossRef]
76. Aljabaa, A.; Almoammar, K.; Aldrees, A.; Huang, G. Effects of Vibrational Devices on Orthodontic Tooth Movement: A Systematic Review. *Am. J. Orthod. Dentofac. Orthop.* **2018**, *154*, 768–779. [CrossRef]
77. Qamar, Z.; Alghamdi, A.M.S.; Bin Haydarah, N.K.; Balateef, A.A.; Alamoudi, A.A.; Abumismar, M.A.; Shivakumar, S.; Cicciù, M.; Minervini, G. Impact of Temporomandibular Disorders on Oral Health-related Quality of Life: A Systematic Review and Meta-analysis. *J. Oral. Rehabil.* **2023**, *50*, 706–714. [CrossRef] [PubMed]
78. Cirulli, N.; Inchingolo, A.D.; Patano, A.; Ceci, S.; Marinelli, G.; Malcangi, G.; Coloccia, G.; Montenegro, V.; Di Pede, C.; Ciocia, A.M.; et al. Innovative Application of Diathermy in Orthodontics: A Case Report. *Int. J. Environ. Res. Public. Health* **2022**, *19*, 7448. [CrossRef]
79. Ballini, A.; Dipalma, G.; Isacco, C.G.; Boccellino, M.; Di Domenico, M.; Santacroce, L.; Nguyễn, K.C.D.; Scacco, S.; Calvani, M.; Boddi, A.; et al. Oral Microbiota and Immune System Crosstalk: A Translational Research. *Biology* **2020**, *9*, 131. [CrossRef]
80. Shaadouh, R.I.; Hajeer, M.Y.; Mahmoud, G.; Murad, R.M.T. Systematic Review: Is High-Energy Laser Therapy (HELT) With Flapless Corticotomy Effective in Accelerating Orthodontic Tooth Movement? *Cureus* **2022**, *14*, e22337. [CrossRef]
81. Erdur, E.A.; Karakaslı, K.; Oncu, E.; Ozturk, B.; Hakkı, S. Effect of Injectable Platelet-Rich Fibrin (i-PRF) on the Rate of Tooth Movement. *Angle Orthod.* **2021**, *91*, 285–292. [CrossRef]
82. Minervini, G.; Franco, R.; Marrapodi, M.M.; Ronsivalle, V.; Shapira, I.; Cicciù, M. Prevalence of Temporomandibular Disorders in Subjects Affected by Parkinson Disease: A Systematic Review and Metanalysis. *J. Oral. Rehabil.* **2023**. [CrossRef] [PubMed]
83. Scarano, A.; Inchingolo, F.; Rapone, B.; Lucchina, A.G.; Qorri, E.; Lorusso, F. Role of Autologous Platelet Gel (APG) in Bone Healing: A Rabbit Study. *Appl. Sci.* **2021**, *11*, 395. [CrossRef]
84. Krishna, V.B.; Duggal, I.; Sharan, J.; Mangaraj, M.; Duggal, R.; Jena, A.K. Effect of Leukocyte-Platelet-Rich Fibrin (L-PRF) on the Rate of Orthodontic Tooth Movement and Expression of Various Biomarkers in Gingival Crevicular Fluid. *Clin. Oral. Invest.* **2023**, *27*, 2311–2319. [CrossRef]
85. Chackartchi, T.; Iezzi, G.; Goldstein, M.; Klinger, A.; Soskolne, A.; Piattelli, A.; Shapira, L. Sinus Floor Augmentation Using Large (1–2 Mm) or Small (0.25–1 Mm) Bovine Bone Mineral Particles: A Prospective, Intra-Individual Controlled Clinical, Micro-Computerized Tomography and Histomorphometric Study. *Clin. Oral. Implant. Res.* **2011**, *22*, 473–480. [CrossRef]
86. Maspero, C.; Abate, A.; Inchingolo, F.; Dolci, C.; Cagetti, M.G.; Tartaglia, G.M. Incidental Finding in Pre-Orthodontic Treatment Radiographs of an Aural Foreign Body: A Case Report. *Children* **2022**, *9*, 421. [CrossRef]
87. Inchingolo, A.D.; Inchingolo, A.M.; Bordea, I.R.; Xhajanka, E.; Romeo, D.M.; Romeo, M.; Zappone, C.M.F.; Malcangi, G.; Scarano, A.; Lorusso, F.; et al. The Effectiveness of Osseodensification Drilling Protocol for Implant Site Osteotomy: A Systematic Review of the Literature and Meta-Analysis. *Materials* **2021**, *14*, 1147. [CrossRef]
88. İşeri, H.; Kişnişci, R.; Bzizi, N.; Tüz, H. Rapid Canine Retraction and Orthodontic Treatment with Dentoalveolar Distraction Osteogenesis. *Am. J. Orthod. Dentofac. Orthop.* **2005**, *127*, 533–541. [CrossRef] [PubMed]
89. Liou, E.J.W.; Huang, C.S. Rapid Canine Retraction through Distraction of the Periodontal Ligament. *Am. J. Orthod. Dentofac. Orthop.* **1998**, *114*, 372–382. [CrossRef] [PubMed]

applied sciences

MDPI

Systematic Review

Mandibular Molar Distalization in Class III Malocclusion: A Systematic Review

Angelo Michele Inchingolo †, Assunta Patano †, Giuseppina Malcangi †, Daniela Azzollini, Claudia Laudadio, Anna Maria Ciocia, Roberta Sardano, Laura Ferrante, Merigrazia Campanelli, Gianna Dipalma *, Daniela Di Venere, Alessio Danilo Inchingolo and Francesco Inchingolo

Department of Interdisciplinary Medicine, School of Medicine, University of Bari "Aldo Moro", 70124 Bari, Italy; angeloinchingolo@gmail.com (A.M.I.); assuntapatano@gmail.com (A.P.); giuseppinamalcangi@libero.it (G.M.); daniela.azzollini93@gmail.com (D.A.); c.lauda@hotmail.it (C.L.); anna.ciocia1@gmail.com (A.M.C.); robertasardano@gmail.com (R.S.); lauraferrante79@virgilio.it (L.F.); merigrazia.92@hotmail.it (M.C.); daniela.divenere@uniba.it (D.D.V.); ad.inchingolo@libero.it (A.D.I.); francesco.inchingolo@uniba.it (F.I.)

* Correspondence: giannadipalma@tiscali.it

† These authors contributed equally to this work.

Abstract: Class III Malocclusion presents a complex orthodontic challenge with various treatment options, including orthodontic camouflage and orthognathic surgery. Among these, mandibular molar distalization stands as an orthodontic approach for treating Class III Malocclusion in adults. This systematic review aims to evaluate the current evidence regarding mandibular molar distalization techniques in the treatment of Class III. The search across PubMed, Scopus, Cochrane, and Web of Science databases yielded 582 articles, from which eight met the inclusion criteria. These inclusion criteria were as follows: English language, full text, studies randomized clinical trials, and retrospective studies that evaluated various methods of mandibular distalization for Class III Malocclusion in adult patients, from 2013 to May 2023. Lower molar distalization has gained attention as a non-surgical alternative with effective and efficient outcomes. However, various treatment modalities have limitations, including reliance upon malocclusion severity, diagnosis, patient cooperation, and operator experience. From the studies analyzed, it was found that techniques using skeletal anchorage with TADs, mini-plates, or ramal plates, offer stable anchorage and controlled tooth movement, they allow unilateral action in cases of asymmetry, and they are the most effective methods for achieving distal body displacement of the tooth. Despite the promising results, the relatively small number of studies calls for more high-quality research to explore the efficacy and outcomes of different mandibular molar distalization approaches. The lack of standardized protocols and guidelines for mandibular molar distalization in Class III Malocclusion is also attributed to the limited available literature.

Keywords: distalization; lower molar; Class III; mandibular prognathism; mini-screw; ramal plate; clear aligners; orthodontics

Citation: Inchingolo, A.M.; Patano, A.; Malcangi, G.; Azzollini, D.; Laudadio, C.; Ciocia, A.M.; Sardano, R.; Ferrante, L.; Campanelli, M.; Dipalma, G.; et al. Mandibular Molar Distalization in Class III Malocclusion: A Systematic Review. *Appl. Sci.* **2023**, *13*, 9337. https://doi.org/10.3390/app13169337

Academic Editor: Bruno Chrcanovic

Received: 13 July 2023
Revised: 4 August 2023
Accepted: 15 August 2023
Published: 17 August 2023

1. Introduction

Third Class Malocclusion (TCM), also known as mesiocclusion, is maxillofacial disorder, characterized by maxillary deficit and/or excess of the mandible, with lower molars positioned mesial to the upper molars and an inverted relationship of the incisors [1].

In adult patients, TCM can involve the basal skeletal when dimensional problems of maxillary bones or dentoalveolar structures occur due to the advanced position of the mandible caused by an occlusal interference [2,3].

The prevalence of TCM differs across different populations, with estimates ranging from 3% to 26% worldwide [4]. This condition poses significant functional, aesthetic, and psychological challenges for affected individuals, making its correction a crucial goal in orthodontic treatment [5–7] (Figure 1).

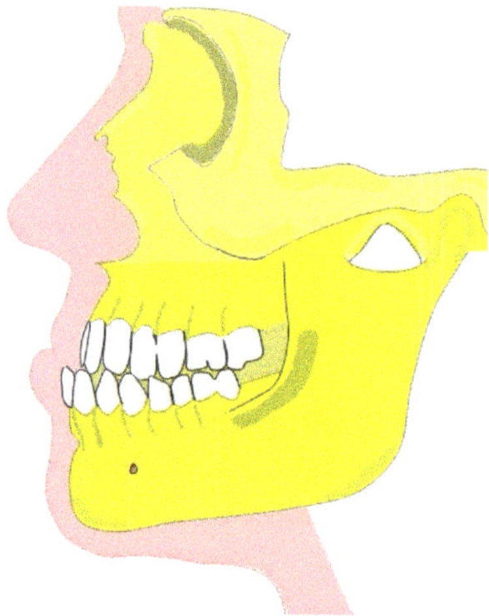

Figure 1. Schematic features of a TCM.

The treatment of TCM traditionally involved either orthodontic camouflage or orthognathic surgery, depending on the severity of the discrepancy [8–10].

Adult patients, who refuse to accept surgical therapy or who are satisfied with their facial appearance, can be treated with dentoalveolar compensation without correcting the underlying skeletal deformity [11–13]. Dentoalveolar compensation treatment offers the non-invasive correction of TCM without surgery, providing aesthetic improvement, enhanced function, and preventive benefits. It retains natural teeth, has a shorter recovery period, and carries a lower risk of complications compared with orthognathic surgery. The dentoalveolar Class III molar relationship can be corrected via differential extractions between the arches (lower arch extractions) or via distalization of the lower molar [14,15].

Mandibular Molar Distalization (MMD) is an orthodontic approach for non-extraction therapy, which aims to gain space in the mandibular arch by moving the mandibular molars distally [16].

The evaluation of mandibular bone quality is of utmost importance in orthodontic treatment planning, particularly when considering MMD. Sufficient space and good bone quality is required to perform MMD. Indications for MMD are TCM, adult patient refusing orthognathic surgery, good patient compliance, absence of third molars. The limitations of MMD concern growing patients, where traditional interceptive treatment is preferred, and patients with severe periodontal disease [17].

By evaluating the quality of the mandibular bone, clinicians can identify potential challenges, develop appropriate treatment plans, and minimize the risk of complications during the distalization process [18,19]. Several studies revealed that the success of this procedure heavily relies on the assessment of the mandibular bone's quality, as it directly influences the stability and efficiency of the distalization process [8,20]. In particular, assessing bone density and the architecture of retromolar trigone helps determine the feasibility of planning MMD. The retromolar trigone is a triangular area located in the jaw, posterior to the last molar, and it is used for the installation of devices that offer an anchoring system for the movement of the lower molars [21].

Radiographic imaging techniques such as panoramic radiographs, lateral cephalograms, and cone-beam computed tomography provide valuable information concerning bone density, root morphology, and cortical thickness. These imaging modalities aid in assessing the quantity and quality of bone, identifying anatomical variations, and limiting factors that may affect the success of MMD [22].

The evaluation of the external oblique ridge, which is observed as the anterior border of the ramus on lateral cephalograms, can be useful for predicting the distalization distance [23]. The cortical contact with the mandibular second molar during MMD negatively influences the process [24]. Orthodontists carefully assess bone quality before starting treatment, as it directly affects treatment planning, mechanics, and appliance selection. Patients with compromised bone quality may require different treatment approaches or additional measures to safely achieve optimal results [22].

For MMD, bone quality must also be adequate because it improves treatment predictability, allowing better anchorage for orthodontic appliances, reducing treatment duration, and ensuring the long-term stability of orthodontic results [25].

This systematic review aims to investigate the various orthodontic methods of performing MMD in adult patients with TCM, and it provides insights into the clinical applicability of MMD. The null hypothesis is that there is no significant difference between the effectiveness of different methods of MMD in the treatment of TCM in adults.

2. Materials and Methods

In this paper, a systematic search was conducted in accordance with the PRISMA 2020 statement [26] to examine the clinical methods used to perform MMD for TCM resolution. The review protocol was registered at PROSPERO under the unique number 443313.

The search string included a combination of keywords related to distal movement and TCM ("Distalization" AND "Class III"). The search, conducted on 11 May 2023, set the time range at 10 years and it specified that the articles needed to be in the English language; the search retrieved 289 articles from PubMed, 88 from Scopus, 131 from Cochrane, and 74 from Web of Science (Table 1).

Table 1. Database search indicators.

Articles screening strategy	Keywords: ("distalization") AND ("Class III")
	Timespan: from January 2013 up to 11 May 2023
	Electronic Databases: PubMed, Scopus, Cochrane, and Web of Science.

Duplicate studies were removed manually. Two researchers independently (D.A., C.L.) performed the selection of articles. Initially, titles and abstracts were reviewed to exclude articles that were clearly not relevant. Then, potentially relevant articles were reviewed in their entirety to assess eligibility in accordance with the inclusion and exclusion criteria, and each discrepancy was resolved by a third researcher (A.P). The following criteria needed to be met by the studies to be included in the systematic review: randomized controlled trials, retrospective studies, studies that had adult patients with TCM as their population, studies in which the main intervention was to use orthodontic appliances to distalize the lower molars, articles that provided data on malocclusion resolution (overjet, cephalometric measurements), and English language publications. Studies with the following characteristics were excluded from the systematic review: studies that were not available as a full text, studies that were not in English, reviews, case reports, articles with patient populations that included children or adolescents, articles that addressed upper molar distalization, MMD achieved via invasive methods combined with orthognathic surgery, and papers that do not provide sufficient or adequate data to evaluate the effectiveness of MMD.

The researchers created a data extraction form, which they then used to extract the relevant data from the eligible studies.

Quality Assessment

Using the Cochrane risk-of-bias tool for randomized trials, Version 2, two reviewers evaluated the articles' bias risk (RoB 2). Any discrepancy was discussed with a third reviewer until an agreement was achieved.

3. Results

The search of the four electronic databases identified a total of 582 studies. After removing duplicates, 468 studies were screened for titles and abstracts. In total, 357 articles were not selected after abstract screening, and 111 were selected for an eligibility assessment. Subsequently, 14 articles were removed because it was not possible to view the full text. Out of the 97 remaining records, 31 articles were removed for research nonconformity, 21 studies focused on upper molar distalization, 5 articles addressed TCM resolution with orthognathic surgery, 11 studies involved underage participants, and 21 articles were excluded because they were case reports. Finally, eight articles were selected for the systematic review. The selection process is summarized in Figure 2.

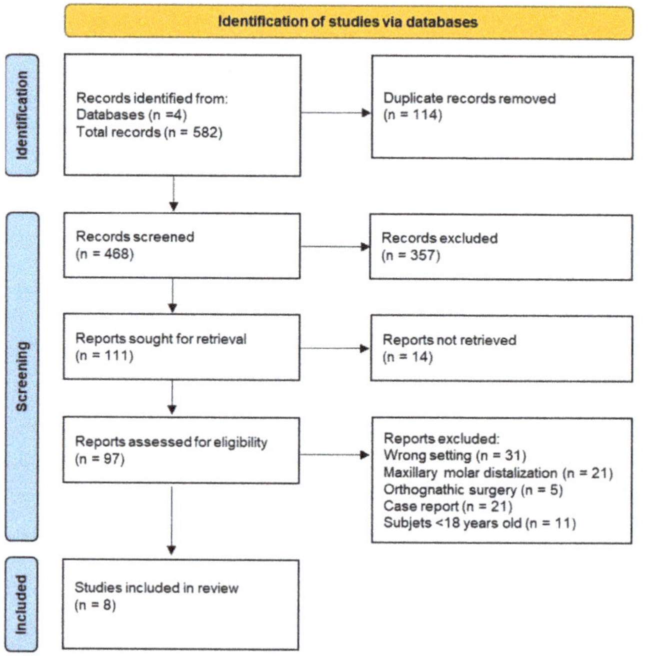

Figure 2. PRISMA flowchart diagram of the inclusion process. The literature search's Preferred Reporting Items for Systematic Reviews and Meta-Analyses (PRISMA) flow diagram.

The articles included are retrospective, and one observational article is included which examines different methods of orthodontic MMD using fixed appliances, clear aligners, and temporary anchorage devices (TADs). The patient samples vary in number and average age, but the common theme concerns the comparison of different distalization methods for the orthodontic treatment of TCM in adult patients.

The data extracted from the studies concerned research characteristics (author year of publication, patient details, distalization method) (Table 2).

Table 2. Study characteristics and results.

Author (Year)	Study Design	Patient (Age)	Distalization Method
He et al. (2022) [27]	Retrospective Study	44 patients (mean age 21 yrs)	Comparison with TADs and fixed appliance
Guo et al. (2020) [28]	Retrospective Study	22 patients (mean age 21 yrs)	Fixed appliance
Nakamura et al. (2017) [29]	Retrospective Study	23 patients (mean age 25 yrs)	Comparison with TADs and fixed appliance

Table 2. *Cont.*

Author (Year)	Study Design	Patient (Age)	Distalization Method
Yeon et al. (2022) [30]	Retrospective Study	40 patients (mean age 26 yrs)	Comparison with TADs and Ramal plates
Azeem et al. (2018) [31]	Retrospective Study	60 patients (mean age 18 yrs)	Comparison with TADs and extraction
Rota et al. (2022) [32]	Retrospective study	16 patients (mean age 25 yrs)	Clear aligners
Yu et al. (2016) [33]	Retrospective study	22 patients (mean age 23 yrs)	Ramal plate and fixed appliance
Ye et al. (2013) [34]	Observational study	19 patients (mean age 20 yrs)	TADs and fixed appliance

Quality Assessment and Risk of Bias

Using RoB 2, the risk of bias was estimated and reported in Figure 3. Regarding the randomization process, 75% of the studies ensured a low risk of bias. However, 75% of the studies excluded performance bias, 75% excluded bias in reported all outcome data, and 25% of the included studies adequately excluded bias in the selection of reported outcomes, whereas 25% excluded bias in self-reported outcomes. Overall, all studies were shown to have a low risk of incurring in bias (Figure 3).

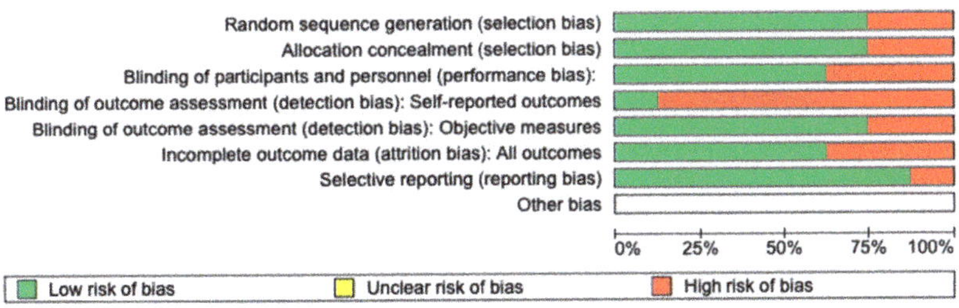

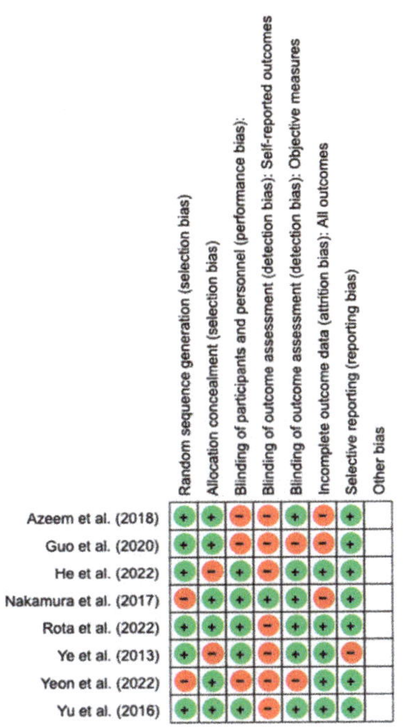

Figure 3. Risk of bias domains of the included studies [27–34].

4. Discussion

The purpose of this systematic review was to gather the most up-to-date scientific evidence available on techniques so that MMD can be used as a nonsurgical treatment for TCM in adult patients. The studies were divided in accordance with the type of method used.

4.1. MMD via a Fixed Appliance

In clinical practice, various mechanics with fixed orthodontic appliances can be used to implement the MMD. The most frequent is the use of intermaxillary elastics. Another technique involves the use of open spiral springs placed between specific brackets, which exert a distal force on the molars [35].

Guo et al. describe the use of the Multiloop Edgewise Arch Wire (MEAW) technique for dental compensation using MMD and the prevention of temporomandibular joint (TMJ) disorders. The MEAW technique uses L-shaped bending forces on the teeth. By applying a distal force to the L-shaped curve of the MEAW arch wire, the lower molars are tilted back towards the distal center and repositioned, helping to correct the anterior crossbite. This approach potentially reduces the risk of TMJ problems such as TMD symptoms or condyle displacement [28]. we inserted the bibliographic reference number next to the names, but unfortunately we could not make the zotero link in the numbers in this image

Hu et al. report a case of MMD with the use of fixed orthodontic appliances, intermaxillary elastics and open coil springs. The treatment resulted in a satisfactory facial profile and stable occlusion [36].

Hisano et al. describe the non-surgical treatment of a TCM with lateral deviation via the distal movement of the mandibular arch and extraction of the third molars. The mandibular molars were moved into an upright position and distalized using a light continuous tip-back force and short Class III elastics [37].

In addition to elastics and springs, modifications to the archwire can be made to incorporate distalizing components. For example, Oliveira et al. use a sliding jig (SJ) combined with intermaxillary elastics. The use of SJ, a type of orthodontic mechanic, allowed for efficient forces to be applied to individual teeth, promoting controlled molar movement. Despite the challenges associated with patient cooperation, the treatment was successful in achieving functional occlusion and dental aesthetics [38].

Nakamura compared the results of orthodontic treatment with TADs and Class III elastics. TADs have been shown to cause the anti-clockwise rotation of the occlusal and mandibular planes, whereas Class III elastics can lead to the clockwise rotation of the mandible. The distalization achieved with TADs results in the body movement of the lower incisors and distal tilting of the lower molars. The amount of distalization achieved with TADs is generally greater than with Class III elastics. They declare that TADs offer a reliable method of distalization for the treatment of TCM [29].

Is important to note that the success of MMD with fixed appliances depends on several factors, including the severity of the malocclusion, the patient's cooperation, and the orthodontist's expertise.

4.2. MMD with Skeletal Anchorage

Skeletal anchorage is highly effective for MMD in orthodontics. It eliminates the need for patient compliance and relies on stable skeletal implants or mini-plates [39]. This approach allows for controlled and predictable tooth movement, while minimizing unwanted side effects on adjacent teeth [25].

4.2.1. TADs

TADs are widely used as stable anchorage in orthodontic treatments in which significant tooth movement is required. TADs in various applications have been demonstrated to optimize orthodontic mechanics, they need very little cooperation from the patient, and they reduce adverse side effects such as mesialization of the anterior anchorage teeth, premolar tipping, molar extrusion, and protrusion of anterior teeth. Advantages of TADs include a decreased treatment period, simple surgical insertion and removal, and a lower risk of potential morbidity [39,40]. Correction of an open bite in a TCM is more challenging when approached nonsurgically, but it is made possible through orthodontic methods with skeletal anchorage [41,42].

Several studies have reported that the total distalization of the mandibular dentition was successful when using skeletal anchorage, and maximizing the distalization of the mandibular arch was seen to reduce the effect of the increased concave profile; this is characteristic of a TCM, given the counterclockwise rotation of the mandible [34,43].

The results of Yan Jin et al. in 2013 also suggest that the application of TADs in the posterior area of the mandible is an effective systematic for nonsurgical TCM treatment [41] (Figure 4A,B).

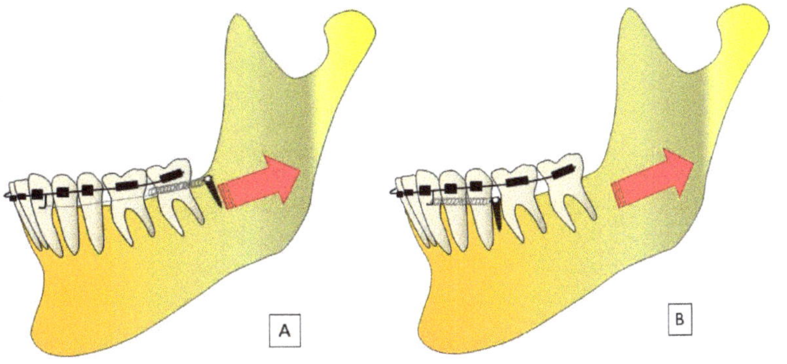

Figure 4. (**A**,**B**) TADSs application sites for MMD.

The application methods of TADs are varied. A study by Z. Zhi-he in 2013 showed that the direct use of mini-screws in the retromolar area required less time and more body movements to retract the mandibular arch without patient cooperation, and this was a better choice for patients with the potential for TJM disorders [34].

Park at al. evaluated cases in which TADs were used for MMD as a function of different force angulations to the mandibular occlusal plane, and for the camouflage treatment of various types of skeletal TCMs. The use of TADs between the roots of premolar and molar, coupled with a thorough knowledge of biomechanics and anatomic limitations, can successfully produce a correct and controlled MMD treatment [44,45].

Generally, two TADs were inserted into the mandibular arch after leveling the plane; one was inserted into the interradicular bone between the mandibular right second premolar and first molar, and the other was inserted between the mandibular left second premolar and first molar. The TADs were placed in different interradicular areas, such as the mandibular left first and second molars, to prevent root damage during installation [46].

Ma et al. used two TADs that were 1.5 mm in diameter in the buccal alveolar bone, one between the mandibular left second premolar and the first molar, and the other between the maxillary right second premolar and the first molar, to provide anchorage in order to correct of a case of TCM subdivision left [47]. TCM treated nonsurgically, with the help of Class III elastics, led to an increase in the angle of the mandibular plane, as compared with the use of TADs which decreased the angle [48]. Class III elastics were preferred for patients with a low angle and short face, whereas TADs were preferred for patients with a high angle and long face [29].

As the direction of the retraction force given to the mini-screws is above the center of resistance of the mandibular arch, the occlusal plane flattens, and the mandibular arch rotates counterclockwise during the distalization movement [49] (Figure 5). As the mini-screws' retraction force is delivered in a direction that is above the mandibular arch's center of resistance, the mandibular arch can be turned counterclockwise when distalized, creating an occlusal plane that is flat [50,51]

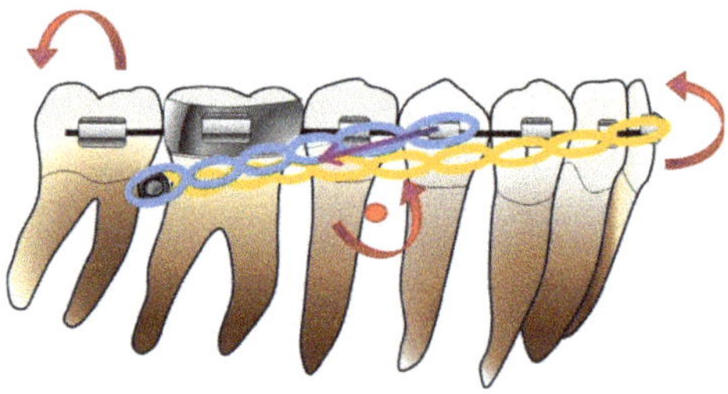

Figure 5. TADs' retraction force.

The goal of a study sponsored by Muhammad Azem and colleagues in 2018 was to compare the effectiveness of a TCM treatment with the removal of two mandibular premolars and the MMD technique. However, regarding the treatment of MMD with skeletal anchorage, it has been seen that the effectiveness of the two treatments when compared is similar in terms of greater patient comfort and less stress [31].

Contextually, regarding the distalization of the lower dental elements, when the closure of a TCM open bite occurs, due to two TADs, applied on the buccal and palatal surface of the upper molars, an intrusive body movement of the maxillary molars is possible with the help of elastic chains passing over the occlusal surface [52]. This system provides excellent control of 'molar inclination during intrusion by evenly distributing the force on the buccal and palatal sides of the elements [41] (Figure 6).

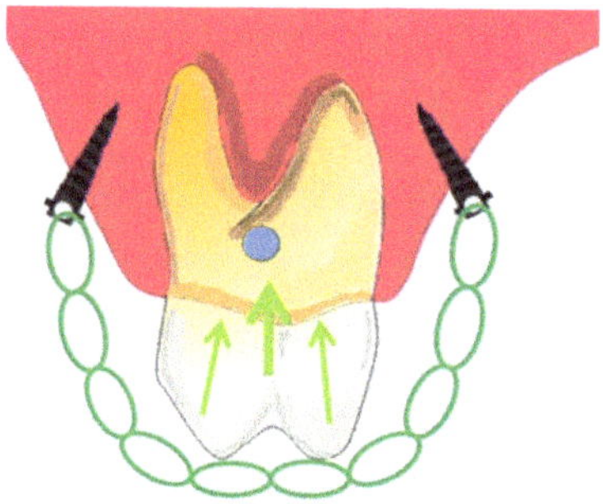

Figure 6. Use of TADs on the vestibular and palatal oral surfaces, for the equal distribution of the intrusive forces that occur with molar intrusions.

A further advantage of TADs is that they can be used monolaterally. In fact, Sha et al. emphasize the use of TADs in cases where the TCM is asymmetric. One of the treatment goals was to correct the mandibular asymmetry via MMD on the affected side. For this purpose, mini-screws were placed in the left retromolar area and between the second lower premolar and the first left molar. These mini-screws provided the anchorage for the distalization of the mandibular dentition. The patient wore Class III elastics to facilitate the desired tooth movement [53].

4.2.2. Mini-Plates

Mini-plates are very stable skeletal anchoring devices because they are held in place by two or more mini-screws [54] (Figure 7). This allows them to be used with heavy forces, and they have been used to good effect when performing distalization of the mandibular arch in adult patients with TCM. As this is a modern approach, there are still few studies and clinical cases in the literature [55]. These include Hakami et al. who published a case report of a TCM treated with the distalization of the lower arch on a mini-plate [56]. The heads of the mini-plates were placed between the first and second molars, and two elastomeric chains, stretched from the canine and first premolar to the mini-plate, on both sides, exerted a force of about 250 g each; these were used to distalize. In this case, a significant MMD of about 4 mm was achieved [56]. In addition, because the force vector passed above the center of resistance of the mandibular arch, a moment was generated that produced the counterclockwise rotation of the anterior teeth, which helped in the correction of the anterior open bite [57].

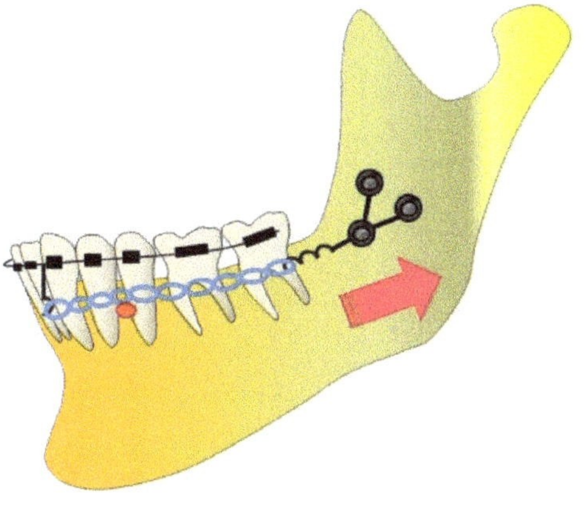

Figure 7. Mini-plates with TADs.

4.2.3. Ramal Plates

The lower arch distalization technique with Ramal Plates (Figure 8) has been proposed in several clinical studies that extoll its advantages.

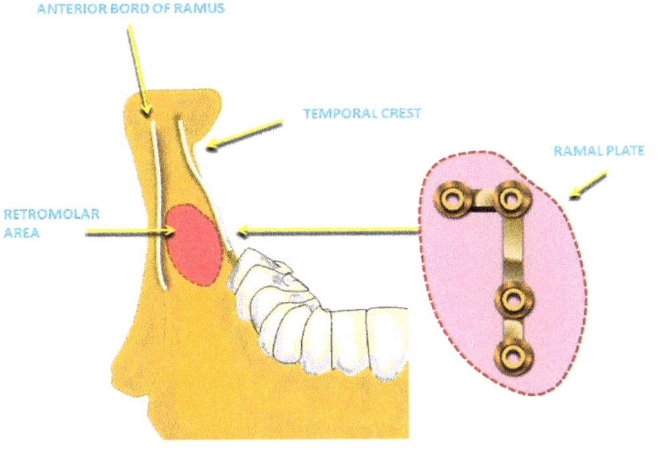

Figure 8. Ramal plate in the retromolar area.

Yoon-Ah Kook et al. published case reports in which they used the ramal plates to perform camouflage in adult patients with TCM who had refused orthodontic treatment combined with surgery [57]. In the case reports proposed by the authors, the insertion technique is described, as follows. The ramal plates were installed in the retromolar fossa, which is located between the anterior border of the mandibular ramus and the temporal crest. A flap was created in the retromolar area, and the L-shaped plate was fixed with two screws, 5 mm long, and 2 mm wide. Then, the flap was sutured so that the anterior hole of the plate was 3 mm lateral to the buccal surface of the second molar so that the force vector was as parallel as possible to the functional occlusal plane [58]. On leveled arches with steel posted arches, an elastic chain was applied between the last ring of the plate and the arch hook. During distalization, it is important to control mandibular anterior tooth movements. As the force vector is above the center of resistance of the anterior teeth, counterclockwise rotation, and thus, uncontrolled tipping of the anterior teeth, must be prevented. To control the position of the root apices, third-order bending in the anterior section of the arch might be useful [57].

The technique proved to be effective and efficient; it allowed the distalization of the entire lower arch in the clinical cases mentioned, which were skeletal TCMs of medium severity, treated without surgery [57,59].

Encouraging results also came from a study by Ye et al., who evaluated the effects of distalization with ramal plates by analyzing the pre- and post-treatment latero-lateral teleradiographs of 22 patients [34]. After the completion of alignment and leveling, 0.019 × 0.025 steel arch elastic tractions were applied and stretched between the last ring of the plate and the bracket of the first molar, exerting 300 g of force per side. The tractions were replaced every 3 weeks, and distalization ended when an acceptable overjet was reached. The valid distalization of the mandibular arch was achieved with ramal plates, without significant changes in the vertical position of the mandibular molars or the angle of the mandibular plane; however, the amount of root distalization was not statistically significant. Therefore, a branch plate can be considered a viable treatment option for the total distalization of the lower arch in TCM patients who refuse surgical treatment or extractive treatments [34].

A clinical study by Byong Moo Yeon et al. made a comparison between distalization performed with ramal plates and distalization performed with mini-screws [30]. The latero-lateral radiographs of 40 patients with TCM, 20 of whom were treated with distalization via mini-screws, and 20 were treated via distalization with ramal plates, were analyzed. It was found that distalization on ramal plates is more effective, and it produces a clockwise rotation of the mandible, unlike distalization with mini-screws which is less effective, but produces greater molar intrusion and counterclockwise rotation of the occlusal plane [34]. Knowing these differences, one can be guided in the clinical choice of treatment, starting with the patient's verticality characteristics [30].

4.3. MMD with Clear Aligners

Currently, there is a shortage of research in the literature regarding MMD performed with aligners, however, there is evidence of the effectiveness of aligners in distalizing the upper molar by more than 2.7 mm with aligners. MMD requires significant force to move the tooth backward, but aligners are primarily designed to apply light and controlled forces for tooth movement [60–62].

In the study by Inchingolo et al., clinicians implemented a compensatory orthodontic treatment for a patient with TCM who had refused orthognathic surgery. The sequential distalization protocol was used, which consists of initially moving most distal teeth while keeping the rest of the arch anchored. Subsequently, the mesial teeth were moved, and the entire arch was repositioned distally (Figure 9). In this way, the distalization is guided tooth by tooth; the patient's cooperation in wearing the templates and the elastics of the third is obviously essential for the resolution of the case [2].

In the study by Rota et al., the effectiveness of the Invisalign system regarding MMD was analyzed. A standardized protocol of sequential MMD, like that used for the upper arch, was also used in this study. The results showed that although the aligner-only method is effective for MMD, the movement that is mainly achieved is not a body movement of the tooth but a tipping movement [32]. This can be explained by the fact that the lower molar has a complex root morphology and greater bone density, which makes it difficult to achieve effective distalization with aligners alone [63].

In a recent study by Auladell, however, the authors described how they planned an inferior distalization movement of more than 3 mm in two TCM patients using aligners and the assistance of mini-implants in the retromolar area. In these patients, the distalization method was proposed to resolve malocclusion and moderate crowding without premolar extractions [64].

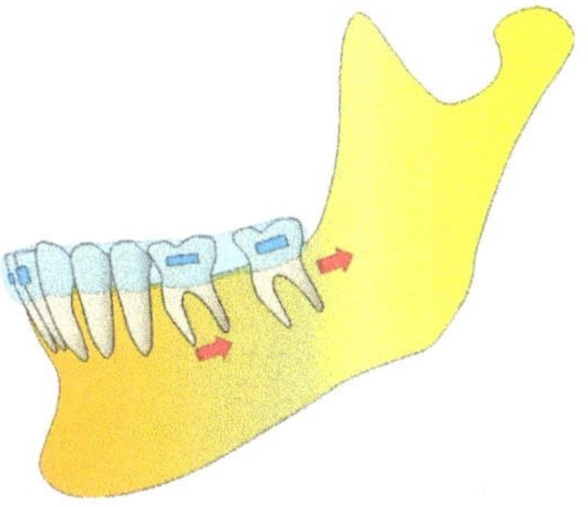

Figure 9. Aligners for the movement of MMD.

The studies highlight the use of sequential distalization protocols and the potential for incorporating mini-implants in the retromolar area to assist with distalization. Further research is needed to explore the efficacy of aligners for MMD and to develop optimal protocols that can achieve successful outcomes in these cases.

Comparing the methods, fixed appliances provide various mechanics with good outcomes, but success relies on patient compliance and orthodontist expertise [36,37]. Skeletal anchorage methods (TADs, mini-plates, and ramal plates) offer stable anchorage, eliminating the need for patient cooperation, and have shown promising results in achieving MMD in various TCM cases [27,29,30]. Clear aligners have limited evidence for MMD but offer benefits like aesthetics and patient comfort; however, their effectiveness in significant distalization remains a challenge [32].

This systematic review may have some limitations, even though MMD seems to be an additional orthodontic strategy for treating TCM in adult patients. First, the validity and representativeness of the results may be impacted by the small number of research papers that are currently available for this topic. Furthermore, the lack of a control group in certain studies makes it challenging to draw clear conclusions regarding the effectiveness of MMD procedures in comparison to other therapies. Furthermore, different research designs and treatment protocols may have been used in the numerous studies included in the review, making it challenging to directly compare and analyze data uniformly. The possibility of publication bias, wherein studies with positive results are more likely to be published than those with negative or insignificant results, as well as the risk of bias and confounding factors in specific studies, a lack of uniformity in reported data, among other issues, could also be limitations. Finally, the variability in patient characteristics, treatments, and evaluation criteria used in the included studies, could affect the generalizability of results for different populations of patients with TCM. Limited data availability may make it difficult to draw firm conclusions and make clinical recommendations based on solid evidence.

Future evidence is needed, such as long-term follow-up studies on the stability of MMD outcomes assessing the degree of recurrence, randomized controlled trials with large samples comparing different techniques to achieve MMD in TCM, and correlation studies with patient-specific factors to identify factors predictive of MMD treatment success.

5. Conclusions

In conclusion, this review sheds light on the potential of MMD for treating TCM in adults. However, it is important to recognize its limitations. Successful MMD depends on factors like malocclusion severity, proper diagnosis, patient cooperation, and the orthodontist's expertise. Patient compliance is crucial, especially for treatments involving elastics, springs, or aligners. Each method has pros and cons. Skeletal anchorage techniques (TADs, mini-plates, or ramus plates) offer stable anchorage and controlled movement, but require further research. Fixed appliances with springs or elastics are non-surgical but rely on patient compliance and have limitations. Aligners mainly provide controlled tipping movements, but they are less effective for significant molar movement. Overall, more research is needed to refine MMD approaches and establish standardized protocols for TCM treatment.

Author Contributions: Conceptualization, D.A., G.M., C.L. and F.I.; methodology, D.A., A.P., R.S., L.F. and G.M.; software, M.C., F.I., D.A. and G.D.; validation, A.M.C., C.L., A.M.C., A.P., A.M.I., D.A., A.D.I., F.I. and G.D.; formal analysis, A.D.I., F.I., R.S., L.F. and G.D.; resources, A.D.I., A.M.I., G.M., A.P., D.D.V., D.A., C.L. and G.D.; data curation, G.M., C.L., A.M.C., F.I. and G.D.; writing—original draft preparation, A.D.I., A.M.I., R.S., A.P., D.A., F.I., G.D. and G.M.; writing—review and editing, D.A., M.C., A.P., D.D.V. and A.D.I.; visualization, A.M.I., F.I., A.D.I., A.M.C., D.A. and A.P.; supervision, A.D.I., A.M.I., A.P., G.M., F.I., G.D. and D.A.; project administration, A.M.I., M.C., C.L., G.D. and G.M. All authors have read and agreed to the published version of the manuscript.

Funding: This research received no external funding.

Institutional Review Board Statement: Not applicable.

Informed Consent Statement: Not applicable.

Data Availability Statement: Not applicable.

Conflicts of Interest: The authors declare no conflict of interest.

Abbreviations

MEAW	Multiloop Edgewise Arch Wire
MMD	Mandibular Molar Distalization
P RISMA	Preferred Reporting Items for Systematic Reviews and Meta-Analyses
SJ	Sliding Jigs
TADs	Temporary Anchorage Devices
TCM	Third Class Malocclusion
TMJ	Temporomandibular Joint

References

1. Ghodasra, R.; Brizuela, M. Orthodontics, Malocclusion. In *StatPearls*; StatPearls Publishing: Treasure Island, FL, USA, 2023.
2. Inchingolo, A.D.; Patano, A.; Coloccia, G.; Ceci, S.; Inchingolo, A.M.; Marinelli, G.; Malcangi, G.; Di Pede, C.; Garibaldi, M.; Ciocia, A.M.; et al. Treatment of Class III Malocclusion and Anterior Crossbite with Aligners: A Case Report. *Medicina* **2022**, *58*, 603. [CrossRef]
3. Dehesa-Santos, A.; Iber-Diaz, P.; Iglesias-Linares, A. Genetic Factors Contributing to Skeletal Class III Malocclusion: A Systematic Review and Meta-Analysis. *Clin. Oral. Investig.* **2021**, *25*, 1587–1612. [CrossRef] [PubMed]
4. Lombardo, G.; Vena, F.; Negri, P.; Pagano, S.; Barilotti, C.; Paglia, L.; Colombo, S.; Orso, M.; Cianetti, S. Worldwide Prevalence of Malocclusion in the Different Stages of Dentition: A Systematic Review and Meta-Analysis. *Eur. J. Paediatr. Dent.* **2020**, *21*, 115–122. [CrossRef]
5. Ngan, P.; Moon, W. Evolution of Class III Treatment in Orthodontics. *Am. J. Orthod. Dentofac. Orthop.* **2015**, *148*, 22–36. [CrossRef] [PubMed]
6. Quinzi, V.; Scibetta, E.T.; Marchetti, E.; Mummolo, S.; Giannì, A.B.; Romano, M.; Beltramini, G.; Marzo, G. Analyze My Face. *J. Biol. Regul. Homeost. Agents* **2018**, *32*, 149–158. [PubMed]
7. Mummolo, S.; Nota, A.; Marchetti, E.; Padricelli, G.; Marzo, G. The 3D Tele Motion Tracking for the Orthodontic Facial Analysis. *Biomed. Res. Int.* **2016**, *2016*, 4932136. [CrossRef]
8. Byloff, F.; Darendeliler, M.A.; Stoff, F. Mandibular Molar Distalization with the Franzulum Appliance. *J. Clin. Orthod.* **2000**, *34*, 518–523. [PubMed]
9. Quinzi, V.; Saccomanno, S.; Manenti, R.J.; Giancaspro, S.; Coceani Paskay, L.; Marzo, G. Efficacy of Rapid Maxillary Expansion with or without Previous Adenotonsillectomy for Pediatric Obstructive Sleep Apnea Syndrome Based on Polysomnographic Data: A Systematic Review and Meta-Analysis. *Appl. Sci.* **2020**, *10*, 6485. [CrossRef]
10. D'Apuzzo, F.; Grassia, V.; Quinzi, V.; Vitale, M.; Marzo, G.; Perillo, L. Paediatric Orthodontics. Part 4: SEC III Protocol in Class III Malocclusion. *Eur. J. Paediatr. Dent.* **2019**, *20*, 330–334. [CrossRef]
11. Kanas, R.J.; Carapezza, L.; Kanas, S.J. Treatment Classification of Class III Malocclusion. *J. Clin. Pediatr. Dent.* **2008**, *33*, 175–185. [CrossRef]
12. Eslami, S.; Faber, J.; Fateh, A.; Sheikholaemmeh, F.; Grassia, V.; Jamilian, A. Treatment Decision in Adult Patients with Class III Malocclusion: Surgery versus Orthodontics. *Prog. Orthod.* **2018**, *19*, 28. [CrossRef] [PubMed]
13. Alhammadi, M.S.; Almashraqi, A.A.; Khadhi, A.H.; Arishi, K.A.; Alamir, A.A.; Beleges, E.M.; Halboub, E. Orthodontic Camouflage versus Orthodontic-Orthognathic Surgical Treatment in Borderline Class III Malocclusion: A Systematic Review. *Clin. Oral. Investig.* **2022**, *26*, 6443–6455. [CrossRef] [PubMed]
14. Bezerra, G.D.C.; Gaschler, J.A.M.; Lourenço, V.S.; Patel, M.P.; Nahás-Scocate, A.C.R.; Maltagliati, L.Á. Compensatory Treatment of a Complex Class III Malocclusion Using Contemporary Mechanics. *J. Clin. Orthod.* **2022**, *56*, 201–209. [PubMed]

15. Rongo, R.; D'Antò, V.; Bucci, R.; Polito, I.; Martina, R.; Michelotti, A. Skeletal and Dental Effects of Class III Orthopaedic Treatment: A Systematic Review and Meta-Analysis. *J. Oral. Rehabil.* **2017**, *44*, 545–562. [CrossRef] [PubMed]
16. Sugawara, J.; Daimaruya, T.; Umemori, M.; Nagasaka, H.; Takahashi, I.; Kawamura, H.; Mitani, H. Distal Movement of Mandibular Molars in Adult Patients with the Skeletal Anchorage System. *Am. J. Orthod. Dentofacial Orthop.* **2004**, *125*, 130–138. [CrossRef]
17. Alogaibi, Y.A.; Al-Fraidi, A.A.; Alhajrasi, M.K.; Alkhathami, S.S.; Hatrom, A.; Afify, A.R. Distalization in Orthodontics: A Review and Case Series. *Case Rep. Dent.* **2021**, *2021*, 8843959. [CrossRef]
18. Özden, S.; Uslu, F.; Dedeoğlu, N. Evaluation of Bone Area in the Posterior Region for Mandibular Molar Distalization in Class I and Class III Patients. *Clin. Oral. Investig.* **2023**, *27*, 2041–2048. [CrossRef]
19. Quinzi, V.; Marchetti, E.; Guerriero, L.; Bosco, F.; Marzo, G.; Mummolo, S. Dentoskeletal Class II Malocclusion: Maxillary Molar Distalization with No-Compliance Fixed Orthodontic Equipment. *Dent. J.* **2020**, *8*, 26. [CrossRef]
20. Kim, H.-J.; Jang, W.-S.; Park, H.-S. Anatomical Limits for Distalization of Lower Posterior Molars with Micro-Implant Anchorage. *J. Clin. Orthod.* **2019**, *53*, 305–313.
21. Wang, Y.; Sun, J.; Shi, Y.; Li, X.; Wang, Z. Buccal Bone Thickness of Posterior Mandible for Microscrews Implantation in Molar Distalization. *Ann. Anat.* **2022**, *244*, 151993. [CrossRef]
22. Nucera, R.; Lo Giudice, A.; Bellocchio, A.M.; Spinuzza, P.; Caprioglio, A.; Perillo, L.; Matarese, G.; Cordasco, G. Bone and Cortical Bone Thickness of Mandibular Buccal Shelf for Mini-Screw Insertion in Adults. *Angle Orthod.* **2017**, *87*, 745–751. [CrossRef] [PubMed]
23. Liu, H.; Wu, X.; Tan, J.; Li, X. Safe Regions of Miniscrew Implantation for Distalization of Mandibular Dentition with CBCT. *Prog. Orthod.* **2019**, *20*, 45. [CrossRef] [PubMed]
24. Choi, Y.T.; Kim, Y.-J.; Yang, K.-S.; Lee, D.-Y. Bone Availability for Mandibular Molar Distalization in Adults with Mandibular Prognathism. *Angle Orthod.* **2018**, *88*, 52–57. [CrossRef]
25. Leo, M.; Cerroni, L.; Pasquantonio, G.; Condò, S.G.; Condò, R. Temporary Anchorage Devices (TADs) in Orthodontics: Review of the Factors That Influence the Clinical Success Rate of the Mini-Implants. *Clin. Ter.* **2016**, *167*, e70–e77. [CrossRef] [PubMed]
26. Liberati, A.; Altman, D.G.; Tetzlaff, J.; Mulrow, C.; Gøtzsche, P.C.; Ioannidis, J.P.A.; Clarke, M.; Devereaux, P.J.; Kleijnen, J.; Moher, D. The PRISMA Statement for Reporting Systematic Reviews and Meta-Analyses of Studies That Evaluate Healthcare Interventions: Explanation and Elaboration. *BMJ* **2009**, *62*, e1–e34. [CrossRef]
27. He, Y.; Wang, Y.; Wang, X.; Wang, J.; Bai, D.; Guo, Y. Nonsurgical Treatment of a Hyperdivergent Skeletal Class III Patient with Mini-Screw–Assisted Mandibular Dentition Distalization and Flattening of the Occlusal Plane. *Angle Orthod.* **2022**, *92*, 287–293. [CrossRef]
28. Guo, Y.; Qiao, X.; Yao, S.; Li, T.; Jiang, N.; Peng, C. CBCT Analysis of Changes in Dental Occlusion and Temporomandibular Joints before and after MEAW Orthotherapy in Patients with Nonlow Angle of Skeletal Class III. *BioMed Res. Int.* **2020**, *2020*, 7238263. [CrossRef]
29. Nakamura, M.; Kawanabe, N.; Kataoka, T.; Murakami, T.; Yamashiro, T.; Kamioka, H. Comparative Evaluation of Treatment Outcomes between Temporary Anchorage Devices and Class III Elastics in Class III Malocclusions. *Am. J. Orthod. Dentofac. Orthop.* **2017**, *151*, 1116–1124. [CrossRef]
30. Yeon, B.M.; Lee, N.-K.; Park, J.H.; Kim, J.M.; Kim, S.-H.; Kook, Y.-A. Comparison of Treatment Effects after Total Mandibular Arch Distalization with Miniscrews vs Ramal Plates in Patients with Class III Malocclusion. *Am. J. Orthod. Dentofac. Orthop.* **2022**, *161*, 529–536. [CrossRef]
31. Azeem, M.; Ul Haq, A.; Ul Hamid, W.; Hayat, M.B.; Khan, D.I.; Ahmed, A.; Khan, M.T. Efficiency of Class III Malocclusion Treatment with 2-Premolar Extraction and Molar Distalization Protocols. *Int. Orthod.* **2018**, *16*, 665–675. [CrossRef]
32. Rota, E.; Parrini, S.; Malekian, K.; Cugliari, G.; Mampieri, G.; Deregibus, A.; Castroflorio, T. Lower Molar Distalization Using Clear Aligners: Bodily Movement or Uprighting? A Preliminary Study. *Appl. Sci.* **2022**, *12*, 7123. [CrossRef]
33. Yu, J.; Park, J.H.; Bayome, M.; Kim, S.; Kook, Y.-A.; Kim, Y.; Kim, C.-H. Treatment Effects of Mandibular Total Arch Distalization Using a Ramal Plate. *Korean J. Orthod.* **2016**, *46*, 212–219. [CrossRef] [PubMed]
34. Ye, C.; Zhihe, Z.; Zhao, Q.; Ye, J. Treatment Effects of Distal Movement of Lower Arch With Miniscrews in the Retromolar Area Compared With Miniscrews in the Posterior Area of the Maxillary. *J. Craniofacial Surg.* **2013**, *24*, 1974–1979. [CrossRef] [PubMed]
35. Quinzi, V.; Ferro, R.; Rizzo, F.A.; Marranzini, E.M.; Federici Canova, F.; Mummolo, S.; Mattei, A.; Marzo, G. The Two by Four Appliance: A Nationwide Cross-Sectional Survey. *Eur. J. Paediatr. Dent.* **2018**, *19*, 145–150. [CrossRef] [PubMed]
36. Hu, H.; Chen, J.; Guo, J.; Li, F.; Liu, Z.; He, S.; Zou, S. Distalization of the Mandibular Dentition of an Adult with a Skeletal Class III Malocclusion. *Am. J. Orthod. Dentofac. Orthop.* **2012**, *142*, 854–862. [CrossRef]
37. Hisano, M.; Chung, C.J.; Soma, K. Nonsurgical Correction of Skeletal Class III Malocclusion with Lateral Shift in an Adult. *Am. J. Orthod. Dentofac. Orthop.* **2007**, *131*, 797–804. [CrossRef]
38. Successful and Stable Orthodontic Camouflage of a Mandibular Asymmetry with Sliding Jigs. Available online: https://journals.sagepub.com/doi/epdf/10.1080/14653125.2018.1444539 (accessed on 21 June 2023).
39. Inchingolo, A.M.; Malcangi, G.; Costa, S.; Fatone, M.C.; Avantario, P.; Campanelli, M.; Piras, F.; Patano, A.; Ferrara, I.; Di Pede, C.; et al. Tooth Complications after Orthodontic Miniscrews Insertion. *Int. J. Environ. Res. Public Health* **2023**, *20*, 1562. [CrossRef]

40. Inchingolo, A.D.; Ferrara, I.; Viapiano, F.; Netti, A.; Campanelli, M.; Buongiorno, S.; Latini, G.; Carpentiere, V.; Ciocia, A.M.; Ceci, S.; et al. Rapid Maxillary Expansion on the Adolescent Patient: Systematic Review and Case Report. *Children* **2022**, *9*, 1046. [CrossRef]
41. Kim, D.; Sung, S. Nonsurgical Correction of a Class III Skeletal Anterior Open-Bite Malocclusion Using Multiple Microscrew Implants and Digital Profile Prediction. *Am. J. Orthod. Dentofac. Orthop.* **2018**, *154*, 283–293. [CrossRef]
42. Rosa, M.; Quinzi, V.; Marzo, G. Paediatric Orthodontics Part 1: Anterior Open Bite in the Mixed Dentition. *Eur. J. Paediatr. Dent.* **2019**, *20*, 80–82. [CrossRef]
43. Aslan, B.I.; Küçükkaraca, E. Nonextraction Treatment of a Class III Malocclusion Case Using Mini-Screw-Assisted Lower Molar Distalization. *Turk. J. Orthod.* **2019**, *32*, 119–124. [CrossRef] [PubMed]
44. Lima, C.E.O.; Lima, M.T.O. Directional Force Treatment for an Adult with Class III Malocclusion and Open Bite. *Am. J. Orthod. Dentofac. Orthop.* **2006**, *129*, 817–824. [CrossRef] [PubMed]
45. Park, J.H.; Heo, S.; Tai, K.; Kojima, Y.; Kook, Y.-A.; Chae, J.-M. Biomechanical Considerations for Total Distalization of the Mandibular Dentition in the Treatment of Class III Malocclusion. *Semin. Orthod.* **2020**, *26*, 148–156. [CrossRef]
46. Tai, K.; Park, J.H.; Tatamiya, M.; Kojima, Y. Distal Movement of the Mandibular Dentition with Temporary Skeletal Anchorage Devices to Correct a Class III Malocclusion. *Am. J. Orthod. Dentofac. Orthop.* **2013**, *144*, 715–725. [CrossRef] [PubMed]
47. Ma, Q.L.; Conley, R.S.; Wu, T.; Li, H. Asymmetric Molar Distalization with Miniscrews to Correct a Severe Unilateral Class III Malocclusion. *Am. J. Orthod. Dentofac. Orthop.* **2016**, *149*, 729–739. [CrossRef]
48. Patano, A.; Inchingolo, A.M.; Laudadio, C.; Azzollini, D.; Marinelli, G.; Ceci, S.; Latini, G.; Rapone, B.; Inchingolo, A.D.; Mancini, A.; et al. Therapeutic Strategies of Primary Molar Infraocclusion: A Systematic Review. *Children* **2023**, *10*, 582. [CrossRef] [PubMed]
49. Patano, A.; Malcangi, G.; Inchingolo, A.D.; Garofoli, G.; De Leonardis, N.; Azzollini, D.; Latini, G.; Mancini, A.; Carpentiere, V.; Laudadio, C.; et al. Mandibular Crowding: Diagnosis and Management—A Scoping Review. *J. Pers. Med.* **2023**, *13*, 774. [CrossRef]
50. Seo, Y.-J.; Park, J.H.; Chang, N.-Y.; Chae, J.-M. Non-Surgical Camouflage Treatment of a Skeletal Class III Patient with Anterior Open Bite and Asymmetry Using Orthodontic Miniscrews and Intermaxillary Elastics. *Appl. Sci.* **2023**, *13*, 4535. [CrossRef]
51. Jing, Y.; Han, X.; Guo, Y.; Li, J.; Bai, D. Nonsurgical Correction of a Class III Malocclusion in an Adult by Miniscrew-Assisted Mandibular Dentition Distalization. *Am. J. Orthod. Dentofac. Orthop.* **2013**, *143*, 877–887. [CrossRef]
52. Laudadio, C.; Inchingolo, A.D.; Malcangi, G.; Limongelli, L.; Marinelli, G.; Coloccia, G.; Montenegro, V.; Patano, A.; Inchingolo, F.; Bordea, I.R.; et al. Management of Anterior Open-Bite in the Deciduous, Mixed and Permanent Dentition Stage: A Descriptive Review. *J. Biol. Regul. Homeost. Agents* **2021**, *35*, 271–281. [CrossRef]
53. Sha, H.N.; Lim, S.Y.; Kwon, S.M.; Cha, J.-Y. Camouflage Treatment for Skeletal Class III Patient with Facial Asymmetry Using Customized Bracket Based on CAD/CAM Virtual Orthodontic System. *Angle Orthod.* **2020**, *90*, 607–618. [CrossRef]
54. Schätzle, M.; Männchen, R.; Zwahlen, M.; Lang, N.P. Survival and Failure Rates of Orthodontic Temporary Anchorage Devices: A Systematic Review. *Clin. Oral. Implant. Res.* **2009**, *20*, 1351–1359. [CrossRef] [PubMed]
55. Sugawara, Y.; Kuroda, S.; Tamamura, N.; Takano-Yamamoto, T. Adult Patient with Mandibular Protrusion and Unstable Occlusion Treated with Titanium Screw Anchorage. *Am. J. Orthod. Dentofac. Orthop.* **2008**, *133*, 102–111. [CrossRef]
56. Hakami, Z.; Chen, P.J.; Ahmida, A.; Janakiraman, N.; Uribe, F. Miniplate-Aided Mandibular Dentition Distalization as a Camouflage Treatment of a Class III Malocclusion in an Adult. *Case Rep. Dent.* **2018**, *2018*, 3542792. [CrossRef] [PubMed]
57. Kook, Y.-A.; Park, J.H.; Bayome, M.; Kim, S.; Han, E.; Kim, C.H. Distalization of the Mandibular Dentition with a Ramal Plate for Skeletal Class III Malocclusion Correction. *Am. J. Orthod. Dentofac. Orthop.* **2016**, *150*, 364–377. [CrossRef] [PubMed]
58. Patano, A.; Cardarelli, F.; Montenegro, V.; Ceci, S.; Inchingolo, A.D.; Semjonova, A.; Palmieri, G.; Pede, C.D.; Mancini, A.; Maggiore, M.E.; et al. Early Functional Orthodontic Treatment of Bad Oral Habits with AMCOP®Bio-Activators. *J. Biol. Regul. Homeost. Agents* **2022**, *36*, 91–110. [CrossRef]
59. Suh, H.-Y.; Lee, S.-J.; Park, H.S. Use of Mini-Implants to Avoid Maxillary Surgery for Class III Mandibular Prognathic Patient: A Long-Term Post-Retention Case. *Korean J. Orthod.* **2014**, *44*, 342–349. [CrossRef]
60. Simon, M.; Keilig, L.; Schwarze, J.; Jung, B.A.; Bourauel, C. Forces and Moments Generated by Removable Thermoplastic Aligners: Incisor Torque, Premolar Derotation, and Molar Distalization. *Am. J. Orthod. Dentofac. Orthop.* **2014**, *145*, 728–736. [CrossRef]
61. Bowman, S.J. Upper-Molar Distalization and the Distal Jet. *J. Clin. Orthod.* **2016**, *50*, 159–169.
62. Pasciuti, E.; Coloccia, G.; Inchingolo, A.D.; Patano, A.; Ceci, S.; Bordea, I.R.; Cardarelli, F.; Di Venere, D.; Inchingolo, F.; Dipalma, G. Deep Bite Treatment with Aligners: A New Protocol. *Appl. Sci.* **2022**, *12*, 6709. [CrossRef]
63. Inchingolo, A.D.; Patano, A.; Coloccia, G.; Ceci, S.; Inchingolo, A.M.; Marinelli, G.; Malcangi, G.; Montenegro, V.; Laudadio, C.; Di Pede, C.; et al. The Efficacy of a New AMCOP(R) Elastodontic Protocol for Orthodontic Interceptive Treatment: A Case Series and Literature Overview. *Int. J. Environ. Res. Public. Health* **2022**, *19*, 988. [CrossRef] [PubMed]
64. Auladell, A.; De La Iglesia, F.; Quevedo, O.; Walter, A.; Puigdollers, A. The Efficiency of Molar Distalization Using Clear Aligners and Mini-Implants: Two Clinical Cases. *Int. Orthod.* **2022**, *20*, 100604. [CrossRef] [PubMed]

MDPI AG
Grosspeteranlage 5
4052 Basel
Switzerland
Tel.: +41 61 683 77 34
www.mdpi.com

Applied Sciences Editorial Office
E-mail: applsci@mdpi.com
www.mdpi.com/journal/applsci

www.ingramcontent.com/pod-product-compliance
Lightning Source LLC
Chambersburg PA
CBHW062033220926
51850CB00010B/2528
9783725822171